T0345005

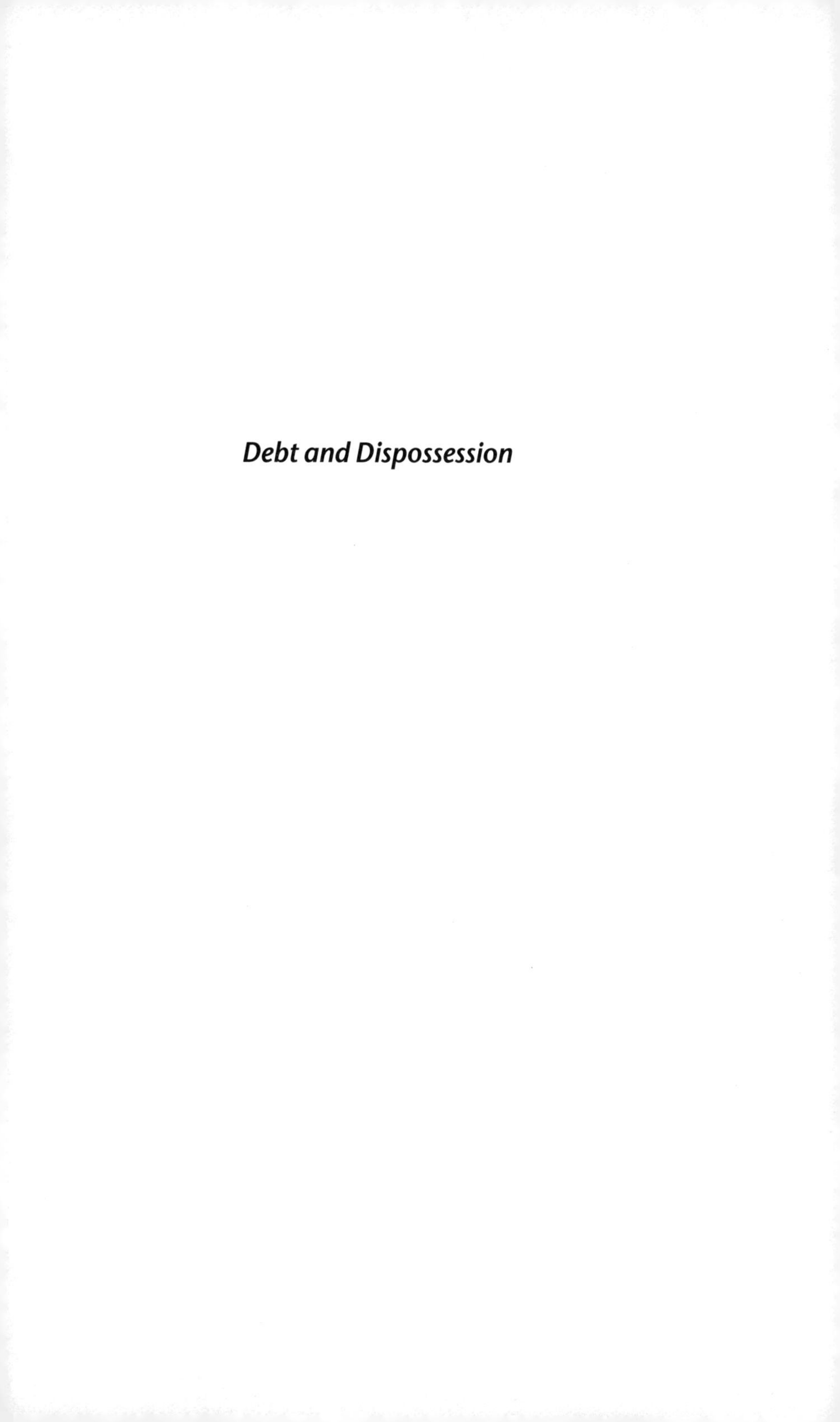

Debt and Dispossession

Debt and

THE UNIVERSITY OF CHICAGO PRESS

Kathryn Marie Dudley

Dispossession

Farm Loss in America's Heartland

CHICAGO AND LONDON

KATHRYN MARIE DUDLEY is associate professor of American Studies and Anthropology at Yale University. Her first book, *The End of the Line: Lost Jobs, New Lives in Postindustrial American* (University of Chicago Press, 1994), is a prize-winning community study of a Chrysler plant closing in Kenosha, Wisconsin.

The University of Chicago Press, Chicago 60637
The University of Chicago Press, Ltd., London

Printed in the United States of America

09 08 07 06 05 04 03 02 01 00 1 2 3 4 5
ISBN: 0-226-16911-1 (cloth)

Library of Congress Cataloging-in-Publication Data

Dudley, Kathryn Marie.
Debt and Dispossession : farm loss in America's heartland / Kathryn Marie Dudley.
p. cm.
Includes index.
ISBN 0-226-16911-1 (alk. paper)
1. Agriculture—Economic aspects—Middle West. 2. Family farms—Middle West. 3. Rural families—Middle West. 4. Farmers—Middle West. 5. Middle West—Rural conditions. I. Title.
HD1773.A3D83 2000
338.1'3'0977—dc21 99-41110

⊗ The paper used in this publication meets the minimum requirements of the American National Standard for Information Sciences—Permanence of Paper for Printed Library Materials, ANSI Z39.48—1992.

IN MEMORY OF SUSAN ROBERTSON

I: "Master, say then

What is this Fortune you mention, that it should have
The world's goods in its grip?" He: "Foolish creatures,
How great an ignorance plagues you. May you receive

My teaching: He who made all of Heaven's features
In His transcendent wisdom gave them guides
So each part shines on all the others, all nature's

Illumination apportioned. So too, for goods
Of worldly splendor He assigned a guide
And minister—she, when time seems proper, spreads

Those vanities from race to race, this blood
Then that, beyond prevention of human wit.
Thus one clan languishes for another's good

According to how her judgment may dictate—
Which is invisible, like a snake in grass.
Your wisdom cannot resist her; in her might

Fortune, like any other god, foresees,
Judges, and rules her appointed realm. No truces
Can stop her turning. Necessity decrees

That she be swift, and so men change their places
In rapid permutation. She is cursed
Too often by those who ought to sing her praises,

Wrongfully blamed and defamed. But she is blest,
And does not hear it; happy among the choir
Of other primal creatures, she too is placed

In bliss, rejoicing as she turns her sphere."

—*The Inferno of Dante,* canto VII,
translation by Robert Pinsky

CONTENTS

PREFACE

The Foelschow farm in western Minnesota was the center of my mother's family. Founded in 1920 by her maternal grandfather, it saw my grandmother Elsie and her siblings through the depression, launching the two sons, Hank and Chuck, in their own hybrid seed corn business and allowing the two daughters, Helen and Elsie, to attend college and become schoolteachers. Vivid memories from my childhood trace back to that farm, a place where everyday events seemed to hold a secret—the tabby cat hunting mice in the corncrib, the snake shedding its skin on the rock by the pasture, and, always, the intoxicating smell of cultivated fields, gathering in the air for the tongue to taste as night fell. Although over thirty years have passed since Hank and his wife, Orline, decided to rent the farm to a neighbor and retire in town, the place has continued to haunt my imagination, now professionally tuned to the mysteries of anthropology and vanishing ways of life. Chuck, Helen, and Elsie are no longer alive; Hank and Orline have recently moved into a nursing home. At times during the research for this book, my quest has been intensely personal. Wading through tall grass and brambles to explore the abandoned homestead on the family property, it seemed all I wished for was a better understanding of the community and culture that shaped my mother's life and, in some way, my own. Why, I wondered, were Hank and Orline the last farmers in our family? And what does this heritage mean to me, a resident of urban areas for most of my adult life?

Debt and Dispossession is a book about how individual families come to terms with the loss of their farm, and how the community as a whole responds to their distress. Because my primary interest is in the "culture" of agriculture—the shared values that give meaning to the events and practices of everyday life—the experience of farm losers is not the only social reality represented in these pages. Equally important to understanding the cultural meaning of farm loss are the voices of those in the community who work closely with farmers and whose jobs, in a very real sense, depend upon them: bankers and other agricultural lenders, farm supply and implement dealers, and state university extension agents. Yet by far the most resonant and culturally dominant voice belongs to farmers who are either in no danger of losing their farm or recent survivors of a close call with financial misfortune. These are the members of the community who set the terms of the local debate about the use of credit and the moral culpability of deeply indebted farmers. Much of the social drama associated with the national farm crisis of the 1980s can be attributed to the schism that opened up between farmers who believed they were in financial trouble through no fault of their own and those who felt that distressed farmers were, on the whole, bad managers who deserved to lose their farm.

I began the fieldwork for this project in 1994. Roughly a decade had passed since the peak years of the farm crisis, and land values in the area were on the rebound. By national measures, the agricultural economy had recovered: debt loads were significantly reduced, and foreclosures had slowed to a trickle. At the community level, however, scars of the crisis remained. In the rural county I chose to study, the farm population had dropped by almost 60 percent from 1970 to 1990. Even so, farming continued to be the major source of employment, providing over 25 percent of the county's total income. To protect the privacy of everyone involved in this research, I have exercised the anthropologist's prerogative and invented pseudonyms for local people as well as local places. Thus, although you will not find "Star Prairie" on a map of western Minnesota, the county's demographic profile (see Appendix) has not been altered and the events I describe are quite real.

Without family ties to the area, it would have been difficult, if not impossible, to conduct the in-depth interviews upon which this book is based. The frank discussion of family and farm finances is not easy under any circumstances, but it can be especially hard to talk about such matters with an outsider, particularly when a painful loss is involved. Of the fifty farm families I interviewed, roughly half lost their farm or came close to losing it during the 1980s. Most of the families in this study had received a letter from the county extension agent introducing me and my project, and were therefore expecting my phone call when I contacted them personally. Nonetheless, lengthy "counter-interviews" often took place, as potential informants pressed me to explain who I was and what I was doing. As doors opened and people welcomed me into their homes, it became apparent that being Hank Foelschow's grandniece was a more appealing feature of my biography than being a professor at Yale. Blood connected me to Star Prairie, even as my "city ways" called to mind native sons and daughters who had left the land.

Interviews with nonfarming members of the community required less personal forms of introduction, for here my academic interest in the transformation of American agriculture was perceived as a point of commonality. In all, I spoke with fifteen professionals who deal regularly with farmers and fifteen individuals who know the community well: longtime residents, clergy, shopkeepers, and local politicians. Usually conducted on the job with no more than one informant, these interviews averaged ninety minutes or less. In contrast, interviews with farm families—which almost always included husband and wife, and occasionally their adult children—took place on the farm and frequently lasted three hours or more. Everyone who participated in this study gave generously of their time, and the hard-earned wisdom they shared with me is the heart and soul of this book.

Beyond formal interviews, I tried to make myself as much of

a presence in the community as possible. I was greatly aided in this effort by my second cousin Ruth Wollan, with whom I lived while conducting interviews. Herself a native of Star Prairie, Ruth had married into one of the county's oldest pioneer families and, in her late sixties, was on a first-name basis with virtually everyone in town. Her delicious meals and delightful companionship made this research possible, and I will always cherish the memory of our summer together. I received fieldwork funding from the Minnesota Historical Society and practical assistance from the University of Minnesota Extension Service. I thank Deborah Miller, Kathleen Mangum, and Jack Morris for all they did to facilitate my research and set me off on the right foot. At Yale University, a Morse Junior Faculty Fellowship granted me a year's leave of absence and the Whitney Humanity Center Research Fund allowed me to hire Katherine Porter, Debra Thurston, and Kristin Bunin to gather U.S. Census materials and transcribe audiotapes of my interviews. Preparing the final manuscript for publication was made infinitely easier with support from Yale's Social Science Faculty Research Fund and the unflagging efforts of Amy Koehlinger and Carrie Lane.

It has been my good fortune to work, once again, with Douglas Mitchell at the University of Chicago Press. His encouragement from the outset, and wise counsel throughout, made every aspect of publication a genuine pleasure. Erin DeWitt's copyediting was exacting in all the right places, and Tom Dodge's photographs do much to bring key moments of the grassroots farm protest to life. Patrick Mooney, David Ostendorf, Peter Marris, and Angelique Haugerud responded with great insight to work-in-progress presented at meetings of the Rural Sociological Society, the Whitney Humanities Center, and the Agrarian Studies Colloquium at Yale. Jane Adams, Douglas Harper, Jean-Christophe Agnew, John Mack Faragher, James Scott, and Kai Erikson offered substantive comments on an early version of the manuscript, while Catherine Wanner, William Peace, and Robert Johnston helped me pull details of the final draft together. Special thanks go to Katherine Newman, my trusted adviser on all things anthropological. Her ethnographic sensibility has long inspired my own, and this book is no exception.

Debt and Dispossession is dedicated to my psychotherapist, Susan Robertson. Had it not been for the many years we devoted to understanding the experience of loss, I would not have had the emotional courage to undertake this research. Her death in the winter of 1997 has made the world an emptier place, even as our work together has helped me appreciate how full that empty space is. Maria Trumpler has lived with this book for longer than any life partner should be expected to. She read every word of every draft—often aloud and many times over—and has spared you, dear reader, from the worst of my excesses. She has my lasting gratitude and deepest love.

How many thousands of times have I seen [my father] in the fields, driving the tractor or the combine, steadily, with certainty, from one end of the field to another. How many thousands of times has this sight aroused in me a distant, amused affection for my father, a feeling of forgiveness when I hadn't consciously been harboring annoyance. It is tempting to feel, at those moments, that what is, is, and what is, is fine. At those moments your spirit is quiet, and that quiet seems achievable by will. But if I look past the buzzing machine monotonously unzipping the crusted soil, at the field itself and the fields around it, I remember that the seemingly stationary fields are always flowing toward one farmer and away from another. The lesson my father might say they prove is that a man gets what he deserves by creating his own good luck.

—Jane Smiley, *A Thousand Acres*

1 Fragile Community

When Dewayne Berg's prayers began to change, his wife knew something was wrong. Early in their marriage they had made it a practice to pray together before starting the day, and Peggy took great comfort in this simple ritual. Always there were prayers for the family, each child mentioned by name, and prayers for those in their church or the community who were ill or troubled. Often Dewayne would ask the Lord's blessing for what he planned to do that day, praying it would be done right and safely. But as the rains began to fall in the spring of 1986, his concerns grew increasingly specific and unusual. He prayed that Peggy would have wisdom when selling the hay, that she would know when to cut it and how to negotiate a good price. These were parts of the farming operation he always handled himself, and Peggy had the uncanny feeling she should be taking notes for future reference. She knew Dewayne was under a lot of stress, what with falling grain prices and the swelling interest payments on their loans, but they had been able to pull out of tough times before. Farming was like that. Just when it seemed like the forecast was as bleak as it could get, crop prices would go up or interest rates would drop, and all would be well again. Yet as weeks went by with no change in sight, Dewayne began to pray as if something terrible were about to happen.

The Bergs had purchased land in southeastern Star Prairie in 1972. Center pivot irrigation was just coming into the area, making it possible for farmers of limited means to purchase land cheaply, invest in an irrigation system, and transform dry, sandy soil into fertile farmland, perfect for light grains and potatoes. Dubbed Bonanza Valley in the 1970s, the sand lands of Star Prairie became a place where you could make a decent living on a family farm, if you were mechanically adept and temperamentally suited to deal with frequent irrigator breakdowns. By 1986, through the gradual acquisition of marginal farmland, the patient clearing of fields, and constant maintenance of their irrigation system, the Bergs were soon farming over a thousand acres, with most of their land in alfalfa. Dewayne, Peggy admits with a smile, loved making hay. The sweet-smelling alfalfa was an ideal crop for this highly erodible soil. Requiring no replanting, it grew from spring to fall, yielding three cuttings—in May, July, late August or early September—and then went dormant through the winter, catching snow and insulating the soil, protecting its fragile root system from the harsh thirty-below winds that blew in from the Dakotas. Each spring the tiny plants would poke through the thawing earth, washing the fields in a misty green, beginning the cycle anew.

That spring, the crops looked good. But Dewayne was preoccupied. He had always walked the fields, usually two or three times a week, even in the most inclement weather, checking the irrigators, monitoring the soil. Peggy liked to join him on these walks, feeling close to him and close to the land. Yet when they walked the fields those last days in May, she sensed something was wrong. When they came upon beaver dams in the slough at the edge of their field, Dewayne flagged the places where beavers had burrowed under the earth to extract mud for their dams, creating a hazard for tractors. Rather than fixing the problem as he would have in the past, he simply marked off the area, as if surrendering part of the land to forces beyond his control. When he placed a circle of flags around a wild duck's nest, concerned the chicks might not hatch before the mowers came, he seemed oddly agitated. Peggy resolved to speak with their pastor to let him know how distressed Dewayne was about their finances and the future. But she never had the chance. Less than a week later, Dewayne was gone.

Peggy's memory of that fateful day remains vivid. For weeks she had been planning a family reunion in Wisconsin for her parents' fiftieth wedding anniversary. On the morning they were to make the drive, Dewayne begged off, claiming a business emergency. With misgivings, Peggy and their three sons left for Wisconsin, hoping he would join them later in the day. That night, when repeated phone calls brought no answer, Peggy began to fear the worst. She called their pastor and asked him to drive out to the farm to see what had happened. Apart from a pile of family photos spread out on the bed, he reported, nothing appeared out of the ordinary. Relieved but apprehensive, Peggy and the boys said hushed good-byes to a concerned family and headed back to western Minnesota. When they got home, the house lights were blazing and a dozen cars were parked along the road.

"All of our friends were here," Peggy recalls, tears welling in her eyes. "They stayed with me all night long."

I try to imagine what it must have been like in the farmhouse that night, here at the large kitchen table where we sit now, almost ten years later, sipping coffee on a bright summer morning. The bustle of well-meaning friends, the warmth of a communal embrace—these sensations come to me easily. Harder is the bone-chilling dread of the unknown. For I know how this story ends. I know that Dewayne survived. After three months as a "missing person," he was located by an officer of the Royal Canadian Mounted Police who pulled him over in the Yukon Territory for driving with illegible mud-splattered license plates. This brush with the law had not brought him home right away, however. Once his whereabouts were discovered, he quit his job and moved on.

Peggy was relieved to learn that her husband was alive, but

the news proved to be a mixed blessing. No longer treated as a potentially grieving widow, she found herself caught in the crossfire of conflicting sentiments that characterized the "farm crisis" of the 1980s. On the one hand, there were those who recoiled at what Dewayne had done and urged her to judge him harshly, if not divorce him outright. On the other hand, there were those who pressured her to declare allegiance with distressed farmers across the country and publicly join their cause. Comfortable with neither course of action, Peggy continued to hope for Dewayne's return, even though this only seemed to alienate her from family and friends:

I never felt he left me or our marriage or the children. I felt he was leaving the farm problems. And I told people that. To some it did not make sense. His brothers were especially harsh on him. Of the people that would have served us papers for foreclosure, they left them on their desks at this point. Because they just felt, maybe partly, that it was bad publicity. At that time there were some very active farm groups around. In fact, some of them were friends. We never participated—but we were sympathetic with them. The Minneapolis paper and the TV people contacted me when Dewayne came out as a missing person. And [the farm activists] said, "Peggy, we could rally around you. We can do this for you. The creditors will back off with this kind of publicity, with you standing there with your children alone." I don't know whether it was wise or unwise, but my decision was that that's not the way I do things, and I don't think that's going to help get Dewayne back. I was afraid that the publicity—if there was a clip somewhere else—that that would make him go further. So I said no. I would not do that.

What most Americans saw of the farm crisis were farmers who chose to make public their private distress. The national tractorcades to Washington, D.C., in 1979 and 1980 are perhaps the best-remembered expressions of discontent in the heartland, but there is an important sense in which these and other collective actions—the "penny auctions" or blocked sales of farm property—did not represent the experience or the politics of most farmers. As Peggy's retreat from the public eye suggests, there were strong disincentives to political involvement, at social as well as psychological levels. Contrary to the impression fostered by the Hollywood movie *Country* and other media events,[1] the majority of farm families faced with foreclosure were not inclined to protest, nor were they universally supported by their communities. Rural sociologists have estimated that less than 2 percent of midwestern farmers took part in public protest during the 1980s, and that fewer than one in a hundred joined political action groups.[2] Moreover, studies of social responses to the crisis have consistently found that distressed farmers report feeling shunned and ostracized by their friends and neighbors.[3]

Hidden from national consciousness are the social and cultural realities of farmers who found the price of protest too exacting or the cost of publicity too much to bear. For every news clip of activists protesting the forced sale of a family farm, tens of thousands of farm families avoided the spotlight, settled out of court, or suffered for years in silence behind closed doors. Economic failure is a stigma in virtually all walks of life, but it is especially discrediting in rural townships, where viable farms remain in the family for generations. A pioneering spirit runs deep in the hearts of those who till the land, and these settlers of the prairie have never looked kindly upon those who succumb to adversity, blame their troubles on others, or start crying for help when the going gets tough. When the names of neighboring families began to appear in the local paper—foreclosures, bankruptcies, auctions, tax delinquencies—there was often little sympathy for the individuals involved and a general consensus that anyone who lost a farm had done something to deserve it. In this context, Dewayne Berg's flight to Canada was only an extreme example of the lengths to which distressed farmers would go to avoid the public admission of failure.

The Pastoral Ideal

The portrait of community that emerges from the story of America's farm crisis is a troubling one, to say the least. We are accustomed to thinking of community life in the rural Midwest as more, not less, socially supportive than in urban areas of the nation. Images of simple folk, earnest neighborliness, and stalwart religiosity come to mind, no matter how dated or fanciful they might be. Even when less desirable elements of rural life enter our awareness, they coexist in an uneasy tension with the pastoral ideal. We may be cognizant of the health risks and environmental hazards posed by modern agriculture—the genetic modification of milk cows; the noxious animal wastes generated by large-scale poultry farms, cattle feedlots, and "hog hotels"; the contamination of groundwater and wildlife as billions of tons of chemical fertilizers, herbicides, and insecticides seep into fields and streams[4]—yet all this is set aside on a Sunday drive through the country. We may know that there are far fewer farmers now than there were decades ago, and that today only about 5 million people—less than 2 percent of the population—live on farms. Yet it rarely occurs to us that over half (60 percent) of all farm families require the wages of off-farm jobs to make ends meet, or that this employment accounts for a significant portion (40 percent) of their income.[5] Moreover, it is easy to dream of the "good life" on picture-perfect farms without realizing that the total family income for a large segment of midwestern farmers (42 percent) is less than $20,000.[6]

There is a serious disconnection between what we know and what we want to believe about farming as a way of life. As cultural historian Leo Marx reminds us, the pastoral ideal has long expressed the "root conflict" of American culture. Where we look to the countryside to find order, beauty, and humane community, the realities of technology and industrial society insistently intrude, reminding us of "the machine's increasing domination of the visible world."[7] The image of "the machine in the garden" is a powerful metaphor for the cultural contradiction between the idea that our lives are shaped by forces beyond our control and the conviction that we are each responsible for our own fate. In the garden, in landscapes of natural splendor, we apprehend the wonder of creation, even as we come into being as individuals and as societies by transforming this pristine world with the machines, technologies, and designs for living that are of our own creation. Only against the backdrop of a universe we *do not control*—be it governed by natural forces or divine will—can the potential for human mastery assert itself.

The pastoral ideal is most often studied as a product of the literary or artistic imagination. The works of poets, novelists, and painters are especially rich sources for scholars interested in how this cultural dilemma has been construed and represented by influential artists at various periods in American history.[8] Much less studied are the ways in which pastoral imagery shapes our expectations, identities, and everyday lives. Yet there is no reason to think that the paradox of human mastery is solely the concern of those engaged in the production and consumption of "high" culture. Particularly in rural settings, where the forces of nature and capitalist society exert direct control over farmers' productivity and material fortunes, the question of what the individual may claim credit or responsibility for is ever present. But the question of individual autonomy—of the degree to which we are able to control our own destiny—also looms large in places insulated from the ravages of hailstorms and commodity markets. Indeed, as modern representatives of the pioneer's epic struggle for survival, family farmers have become our national icon of autonomy.

Not long into my fieldwork in Star Prairie, a Lutheran minister told me a joke about the farmer's exalted independence. We had been talking about how farmers like to think of themselves as stewards of the earth, working in partnership with God to "tend the garden" and feed the world. Inspired by this imagery, the minister recalled a favorite story of his, one about a minister-in-training who went out to a local farm for a visit. Admiring the farmer's fields and the abundant crop, the young minister exclaimed, "Look how God has blessed!" The farmer, feeling insulted, turned to him and replied, "You should have seen it when God had it alone!"

Noticing my puzzled expression, the minister hastened to explain:

The farmer was insulted because he had cleared that land and fertilized it and had this beautiful crop, and before it was just kind of a wasteland. I believe God waits with bated breath for us to use some of the gifts that he has given us—to develop what, by itself, would be waste, into something productive.

Although it might seem odd to think of God's gifts as "waste," this locution captures the local meaning of the pastoral ideal. Land is a "gift" because it is thought to incur a moral obligation on the part of its receivers: a duty to see that it does not to "go to waste" or remain barren and unproductive. The young minister's error lies in praising the Lord's generosity without acknowledging the farmer's fulfillment of a debt or the value of his labor. For in a community based on agriculture as a way of life, it is the figure of the autonomous farmer, cultivating what would otherwise be "wasted" land, who takes center stage in the drama of human mastery. Divine gifts—be they in the form of native talent, good land, livestock, or weather—play an important role, but ultimately it is what the individual does with them that counts. Like the sod-busting pioneers before them, family farmers inscribe their moral character into the landscape, taking pride in what they produce—and in their ability to do it on their own.

Independence and self-reliance are the qualities of character traditionally thought to guarantee success in American society. Not the kind of "success" associated with conspicuous consumption or the excessive display of material wealth, but the moral achievement associated with being "your own boss."[9] It is not for the comforts or pleasures of a moneyed life that the self-reliant strive, but the wherewithal to be beholden to no one. The freedom to call your own shots, take your own risks, and rise or fall on your own merits is, to the independent-minded, the only brass ring worth reaching for. Not only do farmers embody an independent ethos in the kind of work they do, their historic persistence reassures an anxious middle class that this kind of success is still possible in the modern age. To the American worker—subject to mass layoffs, plant closings, corporate downsizings, temporary jobs, and dead-end careers—the family farm glimmers like a promised land, a respectable way to opt out of the rat race, even if that option is never exercised.[10] The imagined autonomy of an agrarian way of life resonates with deeply cherished values in American culture, not the least of which involves the belief that the self-sufficient individual can triumph over all odds. Even those who have not had a farmer in the family for generations can imagine that "returning to the land" is an ultimate test of character and source of spiritual renewal.

The disappearance of a family farm system of agriculture has not yet registered in the consciousness of the nation. The paradox of the pastoral ideal has allowed us to entertain the illusion that any family with the right combination of skill, ambition, and luck can make a decent living on the land. Although it is recognized that farmers of African descent have never fared as well as their European counterparts,[11] this awareness has done little to alter the impression that the iconic "family farm"—a commercial enterprise able to support a heterosexual couple and their children—remains viable in regions of the country where history and geography have favored this kind of farming. Moreover, despite the well-known trend toward larger and fewer farms, the technological changes that drive farmers to expand their operations or quit farming all together are generally seen as signs of progress. "The decline of farming employment is, in many ways, a consequence of success," notes a 1995 U.S. Department of Agriculture bulletin on rural well-being. "Improvements in technology, crop science, and farm management have all boosted output while reducing the need for labor."[12] Taken as evidence of technological mastery, the depopulation of the rural landscape has created a new, ghostly version of the pastoral ideal. At the end of the twentieth century, the machine in the garden has become a portent—not of the powerlessness of traditional lifeways—but of technology's power to mimic tradition itself.

Legacy of the 1930s

When clouds of black dirt appeared on the horizon, shutting out the sun and filling the air with suffocating dust, there was no doubt that the rural Midwest was facing a crisis of epic proportions. Farmers had entered the Great Depression almost a decade before the rest of the nation, as falling commodity prices in the aftermath of World War I reversed a period of rural prosperity that came to be known as the "golden age" of American agriculture. Declining farm incomes were compounded by the collapse of financial markets in 1929, but it was not until the dust bowl swept away hopes of a quick recovery that the magnitude of the disaster sank in.[13] When the unprecedented drought hit Star Prairie in 1934, one farmer recalls, there was genuine fear of widespread starvation:

There was no crop. The binders [machines for binding grain] were not even taken out of the shed. The grain fields were cut with a mower and raked and hauled to the barn for feed. This was mostly Russian thistles, a feed very poor in nutrients but really laxative for the cattle. There was no straw to use for bedding, so it was a real mess in the barn. In May of that year, Dad and I were plowing and in the middle of the afternoon, the wind came up real fast and [carried] a cloud of dust so thick we could not see the place. We unhooked our horses from the plows and started home. After getting the horses tied in the dark, we went to the house. Dust sifted in the windows and doors. The rest of the summer was really dry; it was a real famine.[14]

Although western Minnesotans were spared the worst of the dust bowl, which took its greatest toll in regions to the west and south, the traumatic experience of the 1930s impressed upon an entire generation the virtues of frugality and laying store for an uncertain future. But the lessons of hard times are not restricted to those who live through them. As sociologist Glen Elder Jr. has shown in his study of children born in the 1930s, the Great Depression has provided Americans with an enduring model of what can go wrong with a market society—and how to protect oneself against its inevitable downturns.[15] The notion that the whole economy may "crash" without warning is no longer unimaginable, and the expectation of government intervention created by the Roosevelt administration continues to shape policy to this day. To a society grown accustomed to the idea that individuals create their own good fortune, the depression years came as proof that the economy has a logic, if not a will, of its own.[16]

Yet the legacy of the depression is by no means simple or uniform. Where elders may hew to the wisdom of disciplined saving and conservative spending, the younger generation—facing a different economic milieu—may extol the virtues of credit, even as they point to the 1930s as an example of just how great the risks might be. Anthropologist Katherine Newman has done much to illuminate the "symbolic dialects" that serve to express the divergent concerns of successive generations.[17] To those whose "generational culture" is forged in the crucible of a unique historical event—the depression, say, or the war in Vietnam—the opportunities and dangers of the present will always be open to correspondingly different interpretations. There need not be outright conflict to mark a moment as one defined by competing generational viewpoints, for such differences are often set aside in favor of family unity or upward mobility. Indeed, as Newman demonstrates, intergenerational conflict is likely to be at its most intense—but least overt—when economic realities threaten to derail a family's "class trajectory" or mobility project. At such moments, the elder and younger generations recognize that they must cooperate to achieve a common goal, but opt to pursue it according to strategies that neither agrees upon.[18]

While the problem of intergenerational mobility besets every American family, the conflict can be particularly acute on a family farm. With virtually all of the family's assets tied up in the farming operation, there are hard choices to be made when parents retire and children inherit. Elderly parents must be supported until their death, and nonfarming siblings, as well as those who intend to farm elsewhere, must be paid for their portion of the estate. To the designated heir, coming into possession of a family farm can be like receiving a pie after everyone else has taken a slice. Thus, even if a farm is unencumbered by debt at the time of transfer, most young farmers must take on a mortgage and other farm loans from the outset. Should they hope to support a growing family of their own, the productive capacity of the farm will have to be enlarged and the need for debt financing will be even greater. At such moments, intergenerational tensions often come to the fore. Cautious elders and ambitious youngsters face off across the kitchen table, and in many families, decisions are made that put the farm at risk.

In prosperous times, the transmission of farm assets generally occurs without much fanfare or heartache. Risks may be taken and debts incurred, but confidence in the rising value of farmland tends to still most hand-wringing. One need not have gone through the depression to know that the price of land, like anything else in a market economy, can go down. But farmers recognize that productive farmland is not exactly like other commodities: it is a limited natural resource and, as Star Prairie's farmers like to say, God isn't making any more of it. To farm families seeking to build equity in their operation, land is the soundest investment there is. Throughout the postwar years, farmers could purchase land on credit and watch the value of their holdings rise, or even double, over the period of their mortgage payments. As long as interest rates were low and commodity prices remained steady, the return on a moderate-sized farm (about 180 to 500 acres) generated a decent household income.

The seeds of the farm crisis were sown during the 1970s, a period of low interest rates, rising prices, and unalloyed optimism about the future of American agriculture. Worldwide demand for U.S. farm products was at an all-time high, and the Nixon administration encouraged farmers to plant "fencerow to fencerow," putting more acres into production and, through new chemical means, farming them more intensively. "You never get into financial trouble in hard times," a Star Prairie farmer observes of this period. "It's the good times that cause the problem. You overspend and you become overconfident, then you get into trouble when things get a little tough." By the mid-1980s, get tough it did. International commodity markets slumped and interest rates—spurred by the Federal Reserve's "tight money" policy—soared into the double digits. By the end of the decade, an estimated 200,000 to 300,000 commercial farmers were forced to default on their loans, the majority of these concentrated in the Farm Belt states of Iowa, Minnesota, and Wisconsin.[19] Between 1984 and 1988, 10 percent of all outstanding farm loans were in default, and more agricultural banks failed in 1987 than in any year since the Great Depression.[20]

Surveying the wreckage of the 1980s farm economy, one is tempted to decry the hubris of the agricultural industry, driven by the blind power of technology and deregulated credit markets into a crisis that hindsight tells us could have been prevented. Where were the voices of caution? Was there no one to recall the farm foreclosures and bank failures of the depression? Were there no sober sentinels—no agricultural economists, bank examiners, industry experts, university-trained extension agents, depression-era parents or grandparents—any one of whom, we imagine, should have spotted a disaster on the horizon and sounded a warning? The sad truth, as we shall see, is that such voices were there—but they went unheeded, spoke too softly, or came too late. Like the *Titanic* bearing down upon an iceberg, the operation of the agricultural credit system rested on the assumption that the American farm economy was unsinkable. It would take the failure of one-quarter of a million farms, the collapse of hundreds of rural banks, the government's bailout of the national Farm Credit System, and the demise of a New Deal brainchild—the federal Farmers Home Administration—to prove otherwise.

What Happened to Prairie Populism?

Rural America has historically been fertile ground for third-party politics, and the postwar decline in the farm population has only magnified this tendency. Fewer in number—and increasingly divided by region, farm size, and commodity specialization—farmers rarely conform to the policy agendas or "party lines" drawn for a largely nonrural electorate. After President Jimmy Carter responded to the Soviet Union's invasion of Afghanistan with a grain embargo, even the American Agricultural Movement, the most "radical" farm group at the time, supported Ronald Reagan in 1980.[21] But this "Republican consensus" was rapidly undermined by a deteriorating farm economy and the Reagan administration's attempts to thwart a federally mandated moratorium on foreclosures. Age-old fissures in traditional farm organizations split open, creating spaces for new movements to take hold—and for a paralyzing apathy to set in. Disaffection with the long-standing choice between the "conservative" politics of the Farm Bureau and the "liberal" position of the Farmers Union—or the more "radical" National Farmers Organization (NFO)—led to the spontaneous eruption of myriad protest groups, regional alliances, and national coalitions. What activists shared at this volatile point in rural history was a conviction that the family farm was about to become a casualty of corporate capitalism, and that if it was not saved now, there would be no second chance.

As grassroots protest gained momentum in the mid-1980s, scholars and commentators were inclined to see in the new farm movement a rebirth of the prairie populism that has characterized agrarian revolts on the Great Plains since the late nineteenth century.[22] That activists frequently adopted the tactics and slogans of earlier populist struggles—the "penny auction" of the 1930s Farmers' Holiday movement or the call for "parity" that has accompanied militant holding actions since the New Deal—strongly suggested the revitalization of an older political consciousness, if not a resurgent tradition. Although none of the farm groups openly advocated violence, a series of sensational murders—the deadly ambush of two bankers in Minnesota; the methodical gunning down of a banker, farmwife, and neighboring farmer in Iowa; and the killing of two

U.S. marshals in North Dakota—served to dramatize just how far distressed farmers might go if pushed to the edge. In the pages of the popular press, these horrifying incidents of "madness and murder in the heartland" were evidence of a link between the farm crisis and simmering antigovernment sentiments, and journalists darkly warned that the nation was on the verge of a bloody rural uprising.[23]

Political extremism and populism are not the same thing, of course, but there is a connection. Insofar as a populist ideology portrays the farmer as the victim of economic forces controlled by big business, the government, and international banks, it authorizes a vision of the "little guy" who must use any means necessary to defend home and hearth against a vast conspiracy. Whether the agent of evil is figured as monopoly capitalism, the welfare state, or a cabal of international bankers, populism encourages a "paranoid style" of political thinking in which the individual must forever guard against an enemy capable of many disguises.[24] This reactionary, often racist, impulse is usually downplayed by academics and activists who wish to rehabilitate the populist legacy for a contemporary audience. Thus, while admittedly having a "radical fringe," farm groups of the 1980s fashioned themselves as populists of a New Left variety, emphasizing a "producerist ideology," or what scholars like to call an "incipient class consciousness."[25] In this incarnation, the populist's adversarial stance of "us versus them" pits the "producers"—workers, farmers, and small business owners—against the "parasites"—government officials, corporate executives, and bankers. In theory, the category of "producers" can include anyone who produces a material good or provides a useful service, hence the much-celebrated potential for political "unity" between farm groups and labor unions. In practice, however, these categories are quite elastic—and the sense of solidarity among producers is more imagined than real.

All farmers are engaged in the production of agricultural commodities, but they do not all produce the same products or do the same kind of work. As farming has become more capital-intensive, the differences between farmers who specialize in grain, dairy, poultry, and hog or cattle production have become more pronounced. In years past, a family farm operation typically involved the small-scale combination of several types of farming.[26] Yet as market pressures have pushed farmers to increase their "economies of scale"—that is, to lower the relative costs of production by increasing the quantity produced—farms are becoming highly specialized and tend to concentrate on only one or two types of production. This "balkanization" of agriculture makes it difficult for farmers to think of themselves as belonging to a single interest group, united by common concerns and grievances. Not only is their economic well-being tied to the independent operation of different commodity markets, their work schedules, labor practices, and investment strategies lead them to adopt lifestyles and political positions that are frequently at odds, if not fundamentally antagonistic.

Beneath the invocation of a "producerist" identity lies a complex social reality that is rarely acknowledged in academic and popular accounts of agrarian populism. In addition to being divided by commodity specialization, commercial farmers are, in a very basic way, in competition *with one another* as independent business firms. Agricultural economists have long observed that the historic trend toward larger and fewer farms involves the "cannibalization" of failing farms by more successful ones.[27] The depopulation of rural America is therefore not just a story of how hard it is to "make it" on a family farm. It is also—much less publicly—a story about how hard it is to "make sense" of living in a community where some individuals stand to benefit from the failure of others. For the fact of the matter is that farmers are producers by virtue of their participation in an economic system that rewards competition *between* producers and the "takeover"—hostile or otherwise—of one firm by another. In a community where that firm is also the family home and those competitors are also the next-door neighbors, populist talk of the age-old struggle between small producers and big business leaves a great deal unsaid. Submerged below the public voice of unity and cooperation lies a guarded domain of discourse about farm winners and losers and who deserves to succeed in a capitalist economy.

Political scientist James Scott has alerted us to the importance of distinguishing between public and private expressions of political speech and action.[28] Under conditions of domination where subordinate groups fear that the public declaration of anger or dissent will be met with life-threatening reprisals, resistance may take the form of a "hidden transcript" in which the desire to fight back or "speak truth to power" is acted out in fantasy. Performed or articulated in a specific social site and in the company of a select group of intimates, the hidden transcript allows subordinates to reclaim the dignity that encounters with superiors—the "public transcript"—denies them. Although Scott draws primarily on studies of slavery, serfdom, untouchability, and racial domination, his analysis of hidden transcripts and the "infrapolitics" of subordinate groups offers us a compelling way to think about the political discourse that swirls around the loss of the family farm. For it would be a mistake to assume that farmers' reluctance to engage in public protest during the farm crisis was simply a failure of nerve, on the one hand, or an endorsement of the status quo, on the other. To understand farmers' quiescence or inability to mobilize on a large scale, it is imperative to examine, as Scott suggests, "the social and normative basis for practical forms of resistance."[29] At this level of analysis, we cannot escape the recognition that farmers do not constitute a community unto themselves, but regularly do business and rub shoulders with a wide network of nonfarmers in town and the surrounding urban areas.

Contemporary farming communities may be small and slow paced by metropolitan standards, but they are not socially isolated "islands of history" in which the indigenous culture remains unchanged by contact with the modern world.[30] Rural Americans are linked by transportation and communication systems to the same places and sources of information that the rest of us are, and they participate in the same national debates about education, welfare, and economic opportunity. Indeed, as small-business and home owners, they identify with the broad swath of Americans who consider themselves "middle class" and, as such, often feel exploited by two classes of "parasites"—the very rich and the very poor.[31] But this anxiety is a worldly one, acceptable in conversations about the state of the union in general, but left at the door when it comes to talk of social inequality in their own community. Publicly, a blanket of silence falls over the obvious disparities of wealth that underwrite a banker's "retirement" on a lakeshore resort and a farmer's compulsion to work until he dies "in harness." Nothing is said about an agricultural credit structure that makes it possible for farm families with roots in the community to borrow money from a local bank, while less "connected" farmers must look elsewhere, subjecting themselves to the depersonalizing treatment of distant credit associations and insurance companies or the patronizing supervision of the federal government. At the level of public discourse, denizens of Star Prairie are universally mum about the structures of inequality that give some families an advantage over others.

Yet a counterdiscourse exists at the level of private dialogue, and it is here—in jokes about farm subsidies, gossip about neighbors, and mistrust of lenders—that the cultural subtext of a capitalist economy is revealed. Scratch the surface of communal sociability and you will find lingering resentments, old scores unsettled, and deep suspicion about the motives of others. From the vantage point of the kitchen table, the front porch, or the tractor at the edge of the field, farmers continually monitor the world passing by, reading in the parade of possessions—a new truck, old car, big tractor, or camper with a boat hitch—the moral character of their owners and the standards of consumption in the community as a whole. And they know that they, too, are under constant surveillance. News of their new barn or silo will reach the coffee-shop crowd before the first slab of concrete is laid, while routine purchases at the local co-op or grocery store regularly pass before the eyes of vigilant neighbors. Secretive about their successes as well as their failures, individual farm families live cloistered behind walls of their own making, striving to maintain an appearance of "normality" and fearful of anyone who dares not to conform.

It is within this climate of suspicion that the drama of the farm crisis unfolds. As distressed farmers struggled to hold on to land, dignity, and dreams of a lifetime, they most often did so alone, veiled by a code of silence that could be both protective and suffocating. When the family farm could not be saved, the humiliation of offering up for sale what little was left exposed even the strongest of souls to a mortifying ordeal. For the sacrifice made on the auction block was never merely one of property or material goods. The loss of a farm, as the testimony of dispossessed farmers makes clear, is the loss of a spiritual connection to society and life itself. Because this loss occurs at the hands of their neighbors and friends, collective resistance is difficult, if not impossible. The populist slogans of unity and cooperation, while popular, are to those steeped in the competitive culture of capitalism little more than tunes whistled in the dark. For in the end, farmers realize that they will each be held accountable for their own fate, and they know—as do we all—that social bonds in a market economy are extraordinarily fragile.

TOM Yes, I have tricks in my pocket, I have things up my sleeve. But I am the opposite of a stage magician. He gives you illusion that has the appearance of truth. I give you truth in the pleasant disguise of illusion. To begin with, I turn back time. I reverse it to that quaint period, the thirties, when the huge middle class of America was matriculating in a school for the blind. Their eyes had failed them, or they had failed their eyes, and so they were having their fingers pressed forcibly down on the fiery Braille alphabet of a dissolving economy.

—Tennessee Williams, *The Glass Menagerie*

2 Farm Crisis

Luke Hanson tells a story that makes the old-timers laugh:

On my mother's home place that's been in the family over a hundred years, when I was a senior in high school, my uncle and dad were farming that land, and there was two hundred tillable acres on the farm. We had a hundred-acre field of wheat that we hauled in 1974. Usually at harvest time the prices are a dollar less than during the rest of the year, because they know the farmers have to sell it at that time. We sold wheat for $4.60 a bushel in 1974 and it's never been that high. My dad and uncle paid for that whole farm with that one field of wheat that year! That same year corn prices were $3.40 and they've never been that high since. I told these old-timers, I said, "There's nothing to this farming! I want to get in to this, you know. It would be a pretty easy life. Fun too!" They laughed when I said there's nothing to this.

In his affable way, Luke pokes fun at his youthful credulity as well as at the sense of lost innocence Star Prairie's farmers all share. For only in retrospect is it clear that the high commodity

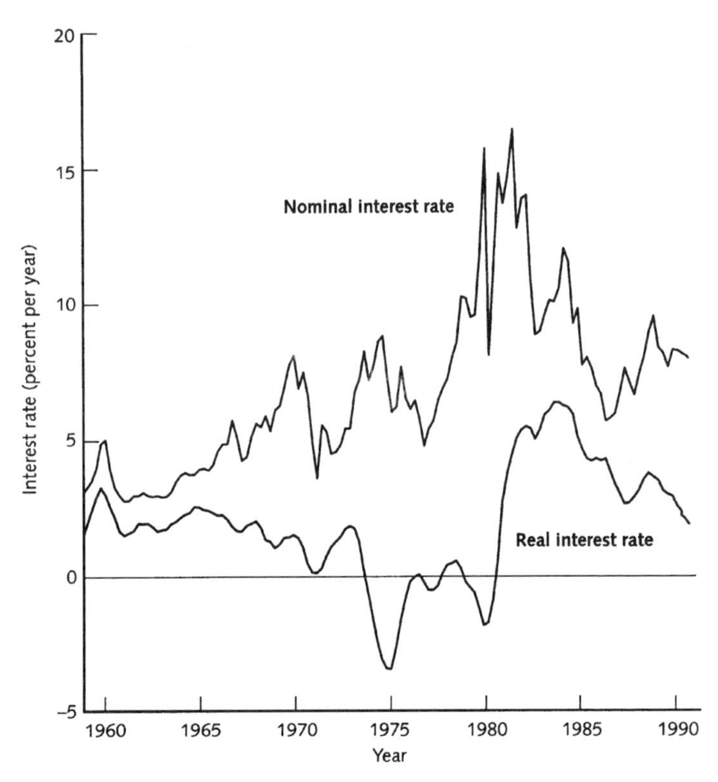

Figure 1 Real versus Nominal Interest Rates
Note: The top line shows the "nominal" interest rate on safe short-term securities (one-year Treasury notes). Most of the upward trend over the last twenty-five years reflects the increase in inflation. The lower line shows the "real" interest rate, equal to the nominal rate minus the inflation rate over the prior year. Real interest rates dropped until 1980, when the Federal Reserve tightened monetary policy.
Source: Federal Reserve Board, U.S. Department of Labor, in *Macro-Economics,* Paul A. Samuelson and William D. Nordhaus, 14th ed. (New York: McGraw Hill, 1992), fig. 10.3, p. 171.

prices of the mid-1970s were a fleeting aberration. At the time, it seemed as though decades of hard work and frugal living were at last going to be rewarded. Exports of American grain were at all-time highs, and domestic credit markets were operating at full tilt. In 1971 President Richard Nixon "unpegged" the dollar, allowing its value to "float" relative to other currencies, making goods produced in this country less expensive to foreign buyers. When poor weather conditions and low crop yields beset a good part of the globe in the early 1970s, demand for U.S. agricultural products surged. The Soviet Union negotiated a multiyear contract for wheat and feed grain in 1972, and within a span of two years, wheat prices doubled and corn prices tripled. Nixon's secretary of agriculture, Earl Butz, called upon American farmers to plant "fencerow to fencerow" and challenged them to "get bigger, get better or get out."[1] The race to feed the world was on, and Star Prairie's young farmers were poised to enter the fray.

Energetic competition would not have been possible, however, were it not for extraordinary credit conditions in the United States. Spurred on by the OPEC oil cartels and soaring energy prices, rapidly escalating inflation served to offset rising interest rates, giving borrowers the unusual luxury of repaying their loans with "cheaper money." During the 1970s, the "real" cost of borrowed money—the nominal rate paid by the borrower minus the rate of inflation—was negligible and, at times, almost "free" (see figure 1). To farmers like Luke Hanson—young, ambitious, and well educated—the time was ripe with possibilities their parents could never have imagined.

Easy Money

Bob Skoglund has been a loan officer with the Farmers Home Administration (FmHA) ever since 1974, when he graduated from college with a degree in economics. Looking back on the 1970s, he describes the period as a "feeding frenzy" caused by loose lending policies:

Interest rates were skyrocketing in the late seventies as people—farmers, consumers, and everybody—was consuming credit like crazy. Rural banks actually ran out of money. We frequently saw some pretty good farmers in our [FmHA] offices, not because the bank didn't want to lend to them, but because they didn't have money to lend to the guy. Inflation was just heating up real good, especially on farmland. [The price of] land was going up. People could borrow against their land to cover their losses, their mistakes, whatever. People that would've been struggling were able to keep going because of the easy credit. People who were doing good used this easy credit to expand, buy machinery, gobble up the neighbor's land, bid up land, whatever. It was a self-perpetuating spiral which kind of fed on itself.

The availability of "easy money" was a great equalizer of financial opportunity. Farmers at all income levels were able to use the rising value of their land as collateral to secure new loans. Community banks had difficulty saying no to valued customers, while insurance companies and the national Farm Credit System instituted newly aggressive lending practices. The benefits of upgrading or expanding a farm operation were an unquestioned article of faith, and agricultural lenders responded to economists' rosy forecasts by enlarging their loan portfolios and actively encouraging farmers to take out new loans.[2] As bank vice president Harvey Beckett observes, there was no reason *not* to lend money to qualified borrowers:

The farmers were making money, and there were incentives for them to go out and buy. They're coming in—and if the lending officer could collateralize the loan and it looked good, based upon the past couple years' profits, it didn't look like any problem with paying it off. So there was a lot of money going out the door. Real estate prices were skyrocketing during that time because farmers had money and they were buying land. Profits were good, so they could pay for it too.

A new golden age for agriculture appeared to be in the offing.[3] Net farm income nearly tripled from 1970 to 1973 and, adjusted for inflation, was 10 percent higher in 1979 than at the start of the decade.[4] To farmers who had watched their parents scrimp and save throughout the 1950s and 1960s—a period when real net farm income had *fallen* by 38 percent—the export boom and expanding money supply heralded a new era.

Steve Schroeder captures the excitement of the moment:

People were running down the halls of Congress saying we were not going to have enough food, and, by gosh, we need to grow food! And, you know, chemicals were coming on board. Sprays. We had one of the first twelve row planters in the whole area, and, gee, that was a huge planter. And, gosh, you know, the push was then for crops. We need cereal commodities to export! Little did we know that it was the low exchange rate of the dollar that was causing the strong exports! [*Laughs.*] And soon we're going to be on the whipping side of that thing! But the dollar was weak, and we could export, and the commodity prices were high. We had three bucks for corn and it was working good. Things worked.

Jake Holquist, a second-generation farmer in Star Prairie, recalls what it was like when he and his brother began farming in the early 1970s:

You just had the big [grain deal]. If you were lucky enough to be farming in 1973—it seems like those years come along periodically. That was the year that Nixon sold all the grain to the Russians. Some call it the Great Grain Robbery. In 1973, you could put the crop in at 1972 prices and that fall, wheat became—it was usually around $1.25 to $2.00 a bushel—that fall, wheat coming out of the field was $5.50. Corn was over $3.00. Farmers are eternal optimists. You're always taught in school that the population is growing and somebody's got to produce all this food and fiber, and your generation's really gonna have it good compared to the generation before. Whether this is a myth or whatever, when you're young, you're ready to take on the mountain. As you get older, the mountain becomes more awesome and more realistic.

Jeannie Travis, a former dairy farmer, puts it this way:

My husband used to quote a little saying: "I want to be a farmer as sure as I have a soul, but how can I learn to be a farmer through this old knothole?" Something about how he was looking over the fence, and his mother would let him at it, if she had a bit of sense. He just knew he wanted to be a farmer. That that was a way to become very filthy rich.

The booming farm economy of the 1970s offered the magic combination of incentives that farmers had been waiting for: high commodity prices and low-cost financing. If the prospect of purchasing farmland with one spectacular crop was what they saw through the knothole, then it was the availability of affordable loans—and the urgings of enthusiastic lenders—that boosted them over the fence.

When Ivor and Helen Lindstrom turned the farm over to their oldest son in 1975, they felt they had timed the transition just right. High commodity prices allowed them to pay off outstanding loans, and they were proud to leave their son a farm unencumbered by debt. Todd had just completed a degree in agricultural economics at the University of Minnesota, and he was eager to put all that he had learned into practice. No sooner did he get his first crop in the ground, however, then the drought of 1976 set him back, as over two hundred acres of corn and soybeans withered under a relentless sun. With poor crop receipts in the fall, he was unable to make payments on the bank loan he had used for a new tractor and operating expenses. Without hesitation, he applied for and received refinancing from the Production Credit Association. Good harvests over the next few years helped him recover from the crop disaster, and he began to draft plans to enlarge his hog operation. When he approached Production Credit with a loan proposal in 1980, he was stunned to discover that they were willing to lend him far more than he was asking for.

HELEN: Sometimes Todd would come home and say, "It just makes me shake, all the money they think I should borrow."

IVOR: In that period of time, it was very difficult to get people to listen to words of caution. I'm sure that there were a lot of older farmers or dads who could remember the depression in the thirties, and they were trying to hold back and say, "Hey, better not do this." But the spirit of the thing at that time was going the other way.

HELEN: Everyone was saying, "They're not making any more

land." It was almost the idea that it didn't make any difference how much you paid for it. They're not making any more of it. There wasn't always a good consideration of what [margin of profit] the land would return.

Among Star Prairie's older generation, the Lindstroms are not alone in observing that their "words of caution" often went unheeded. But they are also aware that their son faced a brave new world—one in which the risks of seizing the moment were as great as those of doing nothing at all. New technology made it possible to farm more acres more cost-effectively, and the purchase of high-priced land could, within reason, be a prudent management strategy. After "retiring" from the farm, Ivor began working as an agricultural counselor at the local technical college. This position helped him appreciate the forces that drove farmers of Todd's generation to pay top dollar for every available acre of farmland. In addition to the "push" they received from lenders hawking easy money, he explains, they were also "pulled" by the lure of new technology:

A farmer considering buying a much larger tractor or combine would say, "We could probably do this." But after investing $50,000 to $60,000 [in the machinery], he would say, "We really need to make this machine work to its fullest extent to get a return on our investment. Therefore, if we can buy this additional eighty acres that our neighbor is going to sell, even though we know we're paying too much for it, we can spread the cost over everything and that way we can use this expensive machinery in a larger way." That was also driving the competition between farmers to buy the neighboring land that was for sale.

By "spreading the cost" of high-priced land, Ivor means that farmers could justify their purchase as a form of "price blending." That is, if they owned 320 acres originally purchased for $200 an acre—a total of $64,000—and then bought 80 acres priced at $400 an acre—a total of $32,000—they could average the price of the new land into the old and say that their 400-acre farm—a total investment of $96,000—cost them an "average" of $240 an acre. If they anticipated that the price of land was only going up, even a steep asking price could appear like a good buy.

A newer, bigger tractor is not just a more efficient way to plow your field. It is also a sign that you are keeping up with the times and improving upon what was done before. New technology, in this sense, is not so much the cause of "progress," but a way of *representing it.* The Lindstrom's youngest son, Jeff, a lawyer in Star Prairie, expresses this feeling well:

Everybody was extremely optimistic. Everybody borrowed a huge amount of money. We didn't have any sense at the time that this was anything but good. We believed that farmers were going to become people of status in this world. That [belief] was there—it was probably a little bit illusory, but you could see it. These people who had been so very conservative in their spending habits in the sixties were now going out—and these young guys were buying new trucks with mags on 'em, you know, and the bigger rollers were buying airplanes and doing these kinds of things. It felt good. It never occurred to me or anyone else to say, "Well, gee, something could go wrong here." It never really occurred to us that land could get to be less valuable again.

So powerful was this sense of forward movement that even debt itself came to represent progress. As Jake Holquist observes:

That was at the time when the exports of U.S. grain was going real well. We were selling grain to the Russians left and right. They were paying for it with gold. And we just thought we were on a roll. I remember one of our loan officers telling us one time, he said, "The one way you can tell if you're doing good is if you owe more this year than you did last year!" We got ourselves heavily in debt like everybody else did. It was quite easy to borrow money. My brother and I still like what they call "fresh paint"—new equipment.

Having a predilection for "fresh paint" isn't a problem if you can pay for it. And during this time, farmers *could* pay for it. In fact, as implement dealer Gary Hendricks recalls, his primary concern was keeping up with his customers' orders, not worrying about their credit-worthiness:

> Everything farmers bought during this period in time, they were able to pay for, write a check for. The problems for us were not having enough equipment, having stuff on order that we had promised the customers that manufacturers didn't supply us on time. At that stage of the game, there was not a financial problem. The financial problem came as [farmers] overexpanded. Some of them became really overextended. But I couldn't see that at the time. All the lenders in the area couldn't see that at the time. Or maybe, the few of them that did couldn't do anything about it, because they had customers going over [to the competitor] and saying, "Well, this bank's going along with it." So it just gets to be a cycle, see?

What are we to think when astute businessmen like Gary Hendricks say that few lenders could see a disaster in the making? Or when well-educated families like the Lindstroms claim that few farmers thought about the dangers of debt? At the level of public discourse, lenders and farmers appear to be in basic agreement: although every loan agreement appeared sound, when seen collectively—as part of a "cycle"—routine financial transactions took on a life of their own. So focused was everyone on the immediate benefits of a particular exchange, this line of reasoning goes, that no one was able to put a stop to the "spirit of the thing" once it had been set into motion. Although this account of the farm crisis passes as the received wisdom in Star Prairie, it papers over critical cultural tensions—not the least of which concerns the issue of who bears the lion's share of responsibility for the fiasco: the farmer or the lender?

From Boom to Bust

If there is an "official" story about the farm crisis—about how the boom of the 1970s turned into the "bust" of the 1980s—it is a story about the march of progress and the triumph of technology over the limitations of human labor, even if the labor "saved" is also, ultimately, the labor lost. Darwinian maxims touting the virtues of competition and the survival of the fittest are invoked to explain the century-long exodus from farming—and to celebrate the superiority of those farmers who remain.[5] To the extent that economic competition in agriculture requires a continual investment in new productive technologies, it has come to necessitate increasingly greater outlays of capital from the outset, as well as throughout the life of the farm. It is with reference to these "realities" that public accounts of the farm crisis take on an aura of inevitability. Who can blame lenders for putting their money to work in such worthwhile ways? And who can blame farmers for trying to become more progressive and competitive? If there is a moral to this story, it is that boom and bust cycles are endemic in the history of capitalism—and, as regrettable as it may be, there will always be some who fail.

Brian Murphy is the University of Minnesota's Agriculture Extension agent in Star Prairie. Hired jointly by the county and the university, his job is to administer agricultural education programs, provide consultation, and oversee outreach services to local residents.[6] During the farm crisis, he was charged with directing the county's state-mandated Farm Credit Mediation Program.[7] In this capacity, he came to know the intimate details of well over a hundred cases of farm foreclosure in the county. Nonetheless, his account of events leading up to the crisis still conforms to the official story:

> Guys like me, ag teachers and others, we kind of got carried away with this whole business of this thing's not ever going to stop. Just buy land and the price will go up! Think of all the money you're going to make—eventually you've got to be making it! And if you get to a situation where you kind of doubt it, sometimes you feel like you're out of step with the rest of the world, you know? Because I always was kind of scared of what might happen. [But] it was almost to the point where people would tell you you're *dumb* if you're not doing this. You know, *everybody else* is doing it.

To be intimidated into silence about an impending catastrophe is a curious state of affairs for someone hired to teach farm management practices, but it draws its credibility from what is by far the most popular way of characterizing the zeitgeist of the 1970s and early 1980s. By this reckoning, the period was dominated by the conviction that bigger is better and debt is good. Doubters and naysayers there may have been, but these contrarians were clearly "out of step" with the times. It is no coincidence that the same philosophy reigned on Wall Street, nor that the era will be remembered as the last heyday of speculative capitalism in the twentieth century.[8] Yet the notion that "everybody is doing it" bears closer examination as a putative rationale for risky behavior. For what could be further from the ideal of the frugal farmer, "bossed" by no one and independent to a fault?

If it were only nonfarmers who blamed the crisis on a herd mentality, we might attribute the authorship of the official narrative solely to them. But the cultural refrain of "everybody is doing it" finds a voice among farmers as well, particularly those who amassed significant debt during this period. Luverne Dahl, a former dairy farmer in his mid-sixties, recalls the sense of pride—and, yes, even independence—that went along with the idea that "everybody" was upgrading and expanding their farms.

LUVERNE: In the seventies, it was fun. Because most of the people that were involved in this, they weren't speculators. They were just plain ordinary folk. It was fun to buy and build. I mean, everybody was doing it. And those that weren't doing it, they were quiet. Maybe there were a few alarmists. But nobody was writing in the paper or putting on TV, "Beware!" It wasn't that way. So those that survived it, they just kept on in their own quiet way. And the rest of us, we were tramping to another tune, another drummer.

KATE: What was that tune?

LUVERNE: When you start out, you start out with your father's machinery. Well, Father started out the same way. Boy, I ain't gonna farm with that junk the old man had. He can have that—but, boy, it ain't for me! So you buy stuff and you have no trouble getting credit, you or anybody else, as long as your net worth went up and you were a good operator.

Luverne acknowledges that, technically, "everybody" was *not doing it,* only those like himself who "tramped to another tune." Yet, to his mind, there is an important sense in which these renegades *were* everybody, since at some point in time, every farmer must strike out on his own. During the 1970s, moreover, virtually every farm family saw their net worth rise as it had never risen before. The value of farmland in Star Prairie had only gone up since the depression, but no one was prepared for the astonishing prices of the 1970s. In 1939 Luverne's father had purchased the home place from his parents for $50 an acre. In 1947 he was able to buy the neighbor's farm at $100 an acre. When Luverne and his wife, Betty, took over the farm in the 1960s, their land was valued at $150 an acre. In 1972, the year before the Russian grain deal, land in the area was going for $200 an acre. After the summer of 1973, however, the price of land increased $100 an acre *every year,* such that by the end of the decade, the Dahls found themselves sitting on land valued at $1,000 an acre. While the price of agricultural land rose rapidly throughout the United States, the run-up was especially steep in Farm Belt states like Minnesota (see figure 2).

With their net worth swelling beneath them, farmers all across the country felt entitled to apply for and receive billions of dollars of credit. To the eyes of accountants, economists, and bank examiners, as Luverne observes, it all looked good—on paper:

But all they were was paper transactions. You can make figures read anything you want to, you know. Well, when your net worth falls, automatically you're in trouble. Then the banker knocks on your door because the examiner starts saying, "Oh, no collateral for this?"

If those swept up in the spirit of the times marched to the beat of a different drum, when the music stopped, they found them-

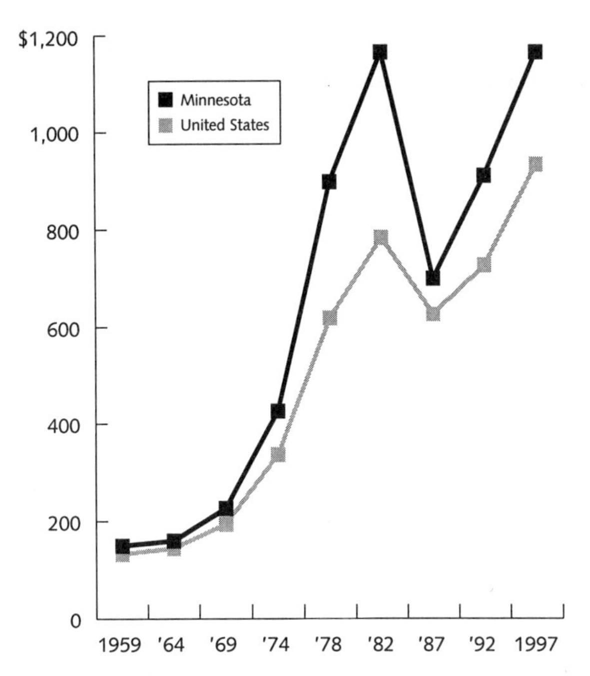

Figure 2 Estimated Market Value of Farmland in the United States and Minnesota, 1959–1997

Note: Data based on a sample of farms.

Source: Bureau of the Census, "Table 1 County Summary Highlights: 1997," *1997 Census of Agriculture,* vol. 1, Geographic Area series.

selves in a surreal game of musical chairs. When commodity prices fell and real interest rates began to rise, farmers scrambled to sit down and secure their financial positions. The problem was not that there were fewer chairs to go around, but that the ones they found collapsed beneath them.

The agricultural boom of the 1970s depended on two trends: worldwide demand for U.S. farm products and low-cost credit. In the 1980s, both came to an end. As countries in Western Europe and South America increased their own agricultural output, America's share of the world market began to shrink and commodity prices fell. In 1979 the Federal Reserve adopted a new monetary policy designed to reduce the nation's annual inflation rate, which at that time exceeded 10 percent. Instead of expanding the money supply to stabilize interest rates, as it had throughout the decade, the Fed began to restrict credit and allow interest rates to rise. Although the policy managed to reduce inflation, it sent "real" interest rates soaring to levels not seen since the Civil War. From an average of 6.8 percent in 1976, the prime rate went to double digits—hitting a record 18.9 percent in 1981.

Although the Fed's restrictive monetary policy made the cost of borrowing money prohibitive for all Americans, the effect on farm families was especially severe. By taking an ever-larger bite out farm incomes, rising interest rates increased the cost of production and reduced the earning capacity of farmland, thereby contributing to the erosion of asset values and the deflationary spiral that gripped the rural economy. Moreover, by driving the value of the dollar higher on foreign markets, stratospheric interest rates made American goods more expensive abroad, further sapping global demand for farm exports.[9] To deeply indebted farmers, it seemed that everything that could go wrong *was* going wrong—all at once. So precariously situated were many, that any number of events could push them into foreclosure. As Steve Schroeder attests, farm failure required a unique mixture of ingredients, none of which would have been fatal alone:

When we bought our farm, interest rates were 7 percent on $600 [an acre of] land; that wasn't too bad. But when the interest rate goes to 13 and 14, that doubles your cost. Plus, then our taxes went from $3 to $9 [an acre]. Seeds, sprays, fertilizer, everything just doubled, doubled, doubled! Everything came together—"Eighty-five," I said, "that's our make-or-break year. We've got to have a crop; we've got to do good this year or we're out of business." It was just tough. It was way too wet. We waited until September and we thought, Oh, we need a good September. The whole year was coming down to one month. In high school when I played football, it was September and it was hot and sweaty! I thought, Boy, if we get a September like that, we'll have a good crop yet! The makings were there. The crops stood there, but it was just too wet. The sun didn't shine. It was dim, dark, gray days. We ended up with very poor crops. It's phenomenal how everything came together to be bad all at once—increasing interest rate, increasing tax, increasing input cost, and poor crops. I never seen a recipe for failure greater than that.

In the view of many farmers, the crisis was magnified by the U.S. government. No sooner had they geared up to fill the world's grain bins than President Carter shut off the pipeline to the Soviet Union in retaliation for its invasion of Afghanistan. While the January 1980 embargo was not the only cause of declining commodity prices, it marked a symbolic end to the proactive trade posture taken by the Nixon administration.[10] To those in agriculture, the embargo added unwarranted insult to injury, coming as it did on the heels of Paul Volcker's October 1979 announcement of the new monetary policy. Steve Schroeder expresses this sentiment well:

What was really kind of bad is that the government controls the economy. They raise the interest rates, and what really hurt is in 1980 they threw that embargo on us. We had our bins full of corn and we were figuring we could export it and get a better price, and all at once they came along and whacked us. Jimmy Carter did that to us, and I'll never forget that day. We woke up and it just shocked us. Now, all at once, our commodities were worth—you know, our profit was gone, because we were counting on that market for the profit. They took all the profit off. So you see that's what can happen, some of those interventions like that? The government doesn't realize that they make people very angry. There's a lot of people that feel abused by the system and we did feel that way.

Elford Hagendorf, a third-generation farmer, agrees:

The biggest problem [in agriculture] I see is that the government jumps in—like, when there's an issue with China over human rights. Suddenly we're going to boycott; we're going to hold back our product. Hell, that's none of their business! We're the ones that are producing the produce, let us be. Fight with them on some other issue. But let us price our product and export our product and let's see what we can do. But every time you get some politician involved in a foreign country, [farmers] suffer. Because we're the people held hostage, or our product is held hostage. Because of some State Department idiot. They have no concept of what farming is in the first place. And yet we're the people that get it taken out on. How often do they banish cars from a foreign country? Or refrigerators? Or computers? But our product gets it every damn time!

Perhaps it goes without saying that foreign policy sanctions would lose their bite if the restricted product were not a basic necessity in short supply. Yet the issue is a central one for farmers, who glean in such events an affront to their claims of independence and respectability. If national policy is made without regard to how it affects farmers, they wonder, then in whose interests is it made? Just who, in other words, is everybody?

Unequal Risks

French anthropologist Herve Varenne has written perceptively of the cultural anxieties that suffuse the way Americans use the concept of "everybody."[11] Eager to deter threats to democratic consensus and fearful of the breakdown of moral order, we call upon the rhetorical unity of "everybody" to restore a fragmented community and "bring it back together." That this strategy works in family and friendship groups, as well as in local and national politics, speaks volumes about the premium that a market society such as ours places on a public sphere devoid of substantive conflict. Nonetheless, it is possible to peer beneath the fissures of what passes as universal accord and apprehend a very different reality.

Like farmers, lenders were also hit by the one-two punch of

tight credit and reduced exports. Frank Tostrud, vice president of the Star Prairie Bank, echoes his customers by placing blame for the farm crisis squarely on the White House steps. Had the Carter administration handled foreign affairs in a manner consistent with the trade goals of Nixon's secretary of agriculture, he argues, the effects of the Federal Reserve's new monetary policy would have been much less severe:

Everything we got in the paper [in the 1970s], the things we got from the Department of Agriculture—from all these smart politicians down there in Washington—was let's produce food here and we'll solve all the country's ills. Consequently, you saw farmland prices go up, saw prices of farm commodities go up. It became very profitable, good business, and we had real inflation out here, particularly on land. During that time, of course, you had this runaway [inflation] when Carter was president. Then they got Paul Volcker in, the chairman of the Fed. In 1979, Paul Volcker said we're going to stop this inflation and we're going to raise interest rates. I agreed with what he was doing. But at that very point in time, it was also when we were feuding with Russia, so Carter stopped the grain sales to Russia. So our commodity prices fell.

When the Fed launched its war against inflation, commercial banks were expected to pass the rising interest rates along to their customers. But as Frank explains, there were limits to what rural banks could do:

In '79 we were charging 8 percent on our farm production loans. In two to three years, we were charging 15 and 16 percent, and we were paying 16 percent for our time money. We did not go up [raise our rates] as much as the metro banks. Actually here—and I think in many rural banks—we shrunk our margin considerably because we just knew the farmers couldn't pay it. When all of a sudden their commodity prices go down and they're paying this kind of interest rate, they were in an impossible situation. Particularly those that obviously had expanded too fast in a manner that wasn't really real prudent.

Despite Frank's professed sensitivity to the financial constraints of his farm customers, a thinly disguised moralism creeps into his retrospective account. What had, only a short time earlier, been a joint process of debt and investment—with lenders aiding and abetting farmers' every move—now gets cast as a problem created by farmers who "expanded too fast" or who were not "real prudent." When the farm economy took a nosedive in the mid-1980s, a certain amnesia fell over the "go-go" years of the 1970s. Who had extended all that credit that was now a mountain of bad debt? What were the incentives that drove lenders to egg farmers on, often encouraging them to borrow more than they asked for? Within the public discourse of "everybody was doing it" allowances are made for the fact that some lenders acted in ways that were not prudent and that both parties to the "farms race" are implicated in the "credit crunch" that followed. But lenders' liability is limited—culturally and economically—by the fact that the risks of debt financing are not equally shared. When push comes to shove, it is the borrower who is held accountable for bad debt and the decisions that incurred it.

Leon Hencks, an adult farm management instructor employed by the Star Prairie public schools, worked closely—as a business adviser and legal advocate—with about two hundred farm families during the years of the crisis. Although he remains sympathetic to the plight of his students, he, too, makes it clear where accountability for failure ultimately lies:

Pointing fingers and finding blame to me is fruitless. There's enough blame to go around. Let's get on with it. Yeah, somebody made mistakes. Yeah, the lenders were guilty of wanting that Farmer Brown to buy that adjoining 160 acres because they had enough equity to do it. And this farmer was paying his debts and paying interest, and the lender was making money on this individual. But in the same way, the lender was not putting a gun to the farmer's head and saying, "Do it! Buy it, or we're not going to support you anymore!" They weren't doing that. Some lenders gave bad advice; some didn't. That farmer maybe felt he was being coerced, but he made his own decision. And then when it came down to losing the farm—rather than blame yourself for making bad decisions, you try to blame somebody else, and they might be partially to blame. But ultimately the decision rests with the person buying the property. Farmers have to take responsibility for that action. We [all] have to be responsible. We have to take accountability for the good decisions we make and especially for the poor decisions we make.

Like "everybody else," Leon concludes, farmers must take responsibility for their decisions—no matter who is to "blame" for the advice given or the mistakes made. Missing from the official account, however, is the fact that the economic conditions of the 1980s did not affect all Americans equally and that farmers themselves were subject to different degrees of hardship depending on their product, farm size, and credit history. Nor is there open acknowledgment of the fact that farm loans often went from "good" to "bad" virtually overnight, not because the farmer failed to repay them, but because the lender was anxious to resecure risky loans by demanding more collateral.

Where farmers had once been treated as "millionaires" due to the rising value of their land, lenders were, by the mid-1980s, left with loans worth little more than the paper they were written on. As the hope of high profits vanished from agriculture, speculators pulled their money out of farmland and went in search of more lucrative investments. As farmers' own purchasing power evaporated, expensive land, machinery, and equipment that had once seemed like "good buys" quickly lost value in the contracting rural economy. Where lenders had once gone to great lengths to extend credit, caution became the watchword, as fears of insolvency gripped the agricultural credit system. New rules for calculating the depreciation of property used as collateral were hastily implemented, and insufficiently secured loans immediately recalled. Few farmers had the ready cash or additional assets to "resecure" their loans, and many received notices of foreclosure not because they were delinquent or in default, but because their loans had grown "larger" than the value of the property securing them.

As forced sales mounted, farmers who had entered the 1980s in relatively good condition began to lose their toehold in an increasingly "slippery" credit structure. Community banks hunkered down and denied credit to all but their best customers, while insurance companies and agencies of the Farm Credit System moved swiftly to purge their portfolios of "nonperforming" farm loans and frequently cut off production credit with little or no warning. The only lender in a position to "catch" the growing number of farmers who were falling through the credit structure was the Farmers Home Administration. In 1982 FmHA enacted a "continuation policy" that loosened credit requirements and allowed the agency to continue to finance existing borrowers, regardless of their ability to repay outstanding debt. The goal was to stem the tide of foreclosures and bankruptcies until economic conditions improved. Thus, the Emergency Agricultural Credit Act of 1984 raised borrowing limits on new FmHA operating loans, extended repayment periods, and increased the availability of subsidized interest rates.[12] This policy initiative served to reduce liquidations, but it also had the unfortunate effect of encouraging distressed farmers to dig themselves even more deeply into debt.

The official account of the farm crisis—that "everybody was doing it" and that only "overextended" farmers got caught—gains its legitimacy largely through the omission of detail. Were these omissions simply in the service of clarity and insight, we might be gratified to learn more of the details, but we could hardly hope to say anything new or add much to our understanding that is not, in a sense, already "known." Indeed, a primary effect of the public story is to discourage curiosity about the details—especially when they reveal social conflict and economic inequality. Yet if what public discourse omits is not simply extraneous or redundant, then we may find in the covert "details" the anthropologist's guiding truth: that there are always other stories to be told—and other worlds to hear from.[13]

BELLIGERENT MAN Is there no one in town aware of social injustice and industrial inequality?

MR. WEBB Oh, yes, everybody is—somethin' terrible. Seems like they spend most of their time talking about who's rich and who's poor.

BELLIGERENT MAN Then why don't they do something about it?

MR. WEBB Well, I dunno. . . . I guess we're all hunting like everybody else for a way the diligent and sensible can rise to the top and the lazy and quarrelsome can sink to the bottom. But it ain't easy to find.

—Thornton Wilder, *Our Town*

3 Moral Economy

Not long into our conversation about the farm crisis, Dick Porter reaches for a yellow legal pad and draws me a picture of "what really happened." At the top of the paper he writes "Real Good Shape" to describe the position of farmers who are able to buy seed and fertilizer with the income earned on last year's crops. He calls these Triple X accounts. He proposes a hypothetical situation, that of a farmer with a feeder cattle operation who goes bankrupt, and on the bottom of the paper, writes "Out of Beef." "No one ever explained this to you?" he asks, as if telling me something I should already know. I shake my head and look on with interest as he draws a long vertical "tube" at the right side of the page. "In the eighties," he says, "as you fell down this tube, if you didn't have a way to get a net under you here or here"—he draws several horizontal lines halfway down the tube—"you went all the way down, and you're out." He adds a large downward-pointing arrow at the end. "When you look back on it," he observes, "it's that simple."

As we discuss the details of Dick and Diane Porter's slide

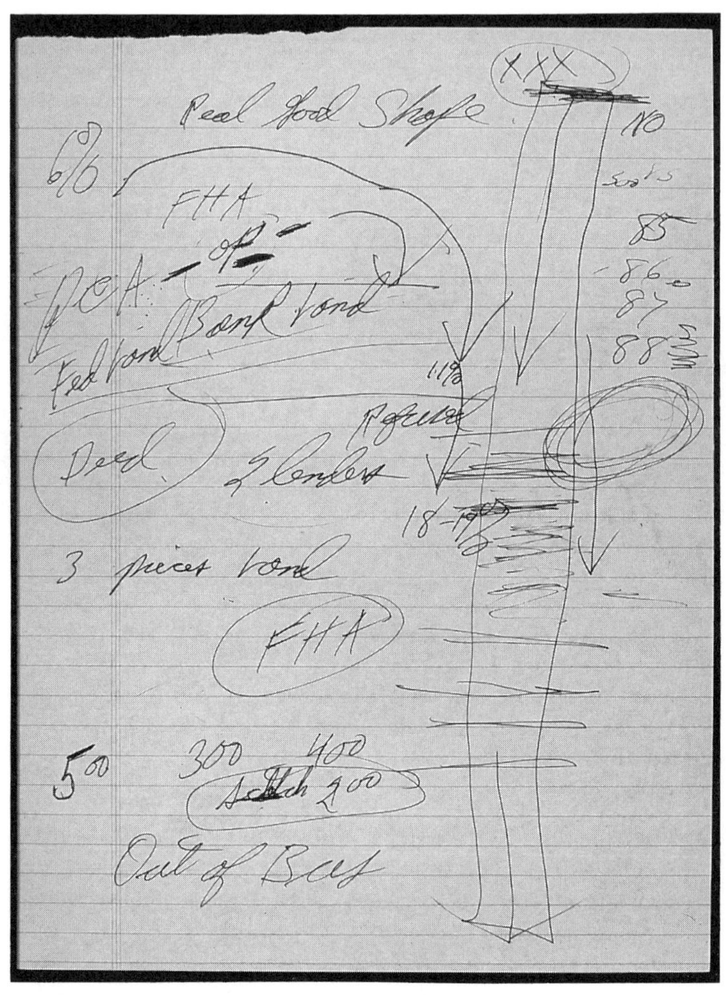

Figure 3 Dick Porter's drawing of the farm crisis.

"down the tubes," the picture fills with dates, interest rates, land values, and more downward-pointing arrows, telling a story that is far from simple. Like other distressed farmers I have talked with—around kitchen tables in the privacy of their own homes—the Porters tell a story in which the farm economy comes to life with perils and pitfalls, and their own experience takes on the quality of a pilgrimage through Dante's inferno. Unlike the public account of the farm crisis, this narrative maps out a social landscape of higher and lower elevations and suffuses it with moral meaning. Winners and losers there may be, but who *deserves* to be one or the other is presented as a drama of individual character and social justice, not simply a "fact" of life in a capitalist economy. Although farmers and lenders have different versions of this "hidden transcript,"[1] both have far more to say about the moral order of the American economy than they ever acknowledge publicly.

The Credit Structure

The Porters' story unfolds as a harrowing descent into hell. Their troubles began in 1982 and 1983, when rising interest rates and poor crops put them behind on their Farm Credit loan for operating expenses. Needing a new loan for spring planting in 1984, they negotiated a new loan package with Farm Credit, one that consolidated their outstanding debt with the mortgage on their land. When they had purchased land in the 1970s, their interest rate was 6 percent. When they refinanced their mortgage in 1984, it had gone up to 11 percent—making the loan consolidation an expensive proposition. Even so, Dick believes, they could have met their obligation if interest rates had not continued to rise, hitting highs of 18 and 19 percent in 1985. Dick calculates that their annual interest payment in 1985 averaged $1,000 *a week*—an amount that quickly became more than they could handle. Determined to continue farming, they sought out the "safety net" of the Farmers Home Administration, the so-called "lender of last resort." Farmers Home took out a second mortgage on their land and issued new operating loans at subsidized interest rates.

Although appreciative of their new lease on life, the Porters despair of ever being able to work their way out of debt. After loan payments, the farm just barely "breaks even," and they rely almost entirely on Diane's salary as an elementary school teacher to meet their living expenses. This would be bad enough, Dick says, but what really gets him down is what it's like to be an FmHA borrower. The way he is treated, he fumes, you'd think he was an incompetent ne'er-do-well. He can't even walk into the local co-op to buy seeds, fertilizer, or fuel without prior approval from Farmers Home. This despite the fact that he has never once defaulted on a loan or stuck a local merchant with an unpaid bill. What angers him is not that the rules of the game are unfair or inherently exploitative, but that they are not applied evenly across the board. The fact that some farmers were able to get "write-downs" reducing the total amount they owe on their loans is, for Dick, a perfect example of this injustice. If he had been able to get this kind of "break," he insists, he could be carrying his debt burden with no problem. Now he is in the untenable position, he complains, of trying to compete with farmers who were given an advantage he was denied.

DICK: I can't believe nobody's explained this to you [referring to his drawing].

KATE: Not in the same way, but certainly with the same feeling of falling into a hole that's hard to get out of.

DICK: I'm not talking feelings here. This is more of a factual scenario. Yeah, *feelings*—I'm frustrated because some of these guys that are farming two thousand acres should be driving a delivery truck, and then you'd be back on a level playing field! Because they're doing the things that the guys that are up here [Triple Xs] are doing because they've had some really big breaks.

In Dick's distinction between "fact" and "feeling," I hear the tension between public and private accounts of the farm crisis. On the one hand, he wants to make it clear that his account is a "factual scenario"—that he *knows* what caused his problems. He is surprised that I don't "know" this too, for, as he believes, it is a matter of public knowledge. This is the story that everyone in Star Prairie knows to be "true": the official story of how basic economic forces—interest rates and commodity prices—affected "everybody" in the same way, especially those who began the decade in debt. On the other hand, Dick acknowledges that his "feelings" *about* this factual scenario are not necessarily shared—at least not by those purported to have had "some really big breaks." The specter of delivery truck drivers masquerading as top-tier farmers suggests a world that has been turned topsy-turvy, with the undeserving on top and the hardworking on the bottom. Poking up through the public story are feelings of anger and ambivalence, humiliation and betrayal—and a haunting sense that something has gone seriously wrong in America.

Steve and Jim Schroeder farmed together for six years after their father died in 1965. When Steve got married in 1971, he felt the time was right to strike out on his own, so he sold his share of the farm to Jim, who remained on the home place. "We didn't have any trouble doing that," Steve says, "largely because the credit was available." The Schroeder brothers were prime candidates for private-sector financing, having inherited a farm owned "free and clear." Jim was able to leverage the home farm to buy Steve's share of the partnership, while Steve was able to use this money as a down payment on nine hundred acres several miles to the north. Steve and his wife, Kathy, had no difficulty obtaining a mortgage on their land from a life insurance company and a loan for machinery and operating expenses from the local bank. Nonetheless, they were not free to do exactly as they wished:

It's easier to get started in land, in a way, because everything [you need] is mobile. In other words, you buy a cultivator—and if you don't pay for it, they just take the cultivator. If you buy a tractor, they'll just take the tractor. So being young people, we could get started in the land, and the credit was there—easier than if we got into any livestock. Because there weren't any buildings [on our land] and to get a livestock operation going, it takes buildings, it takes concrete. Things that a bank can't easily recover. So that's why we were kind of forced into going the route that we did.

Where there was credit, we could go. I think that happened a lot. There were lots of people that simply went the route of where there was credit, they could go. The road map, if you see what I'm saying. That's what we did.

To say that credit is the "road map" is to acknowledge that, however clear you are about your destination, there will be detours and tolls and roundabout ways of getting there. But not all travelers on the highways and byways of credit are subject to the same sense of irreversible consequences that farmers are. When we receive a home mortgage, a car loan, or educational financing, we are usually taking on debt after we have decided upon the house, car, or college of our choice. Farmers have access to credit for these kinds of consumer purchases, but their business decisions are often made in consultation with their lenders and depend on the type of credit they are able to get.

Agricultural lenders fall into two basic categories: those in the *private* sector; and those in the *public* sector, who offer credit subsidized in part or in full by the federal government. Within these categories, there are three types of loans: long-term mortgages on *land,* and short-term loans for *chattel* (that is, livestock, machinery, and equipment) or *operating* expenses (seed, feed, fertilizer, chemicals, and so on). Long- and short-term loans are available in both the private and public sectors, but some lenders specialize in particular types of loans (see table 1).

Steve and Kathy Schroeder were able to set up a farming operation that is relatively large by Star Prairie's standards, where the average farm is around three hundred acres. Community banks tend to be gun-shy of large land mortgages, especially when the buyer is a beginning farmer with few assets. For if there is a lending institution with a living "memory" of the depression, it is the independently-owned rural bank. Local bankers (and their board of directors) still have respect for the grandfatherly advice about avoiding real estate loans, since interests rates and land values fluctuate over the long haul and these notes cannot be "called in" during an economic panic.

Table 1 Forms of Agricultural Credit

Lender	Type
Private	
Family members	Inheritance
Family members, neighbors	Land (contract-for-deed)
Voluntary associations	Land, chattel, operating
Community banks	Land, chattel, operating
Merchants, suppliers, dealers	Chattel, operating
Insurance companies	Land
Private/Public	
Farm Credit Services	
Federal Land Bank	Land
Production Credit Association	Chattel, operating
Public	
Farmers Home Administration	Land, chattel, operating, and loan guarantees

Over the postwar period, life insurance companies were the major providers of institutional farm mortgages, and the Schroeders were ideal candidates for a loan of this type.[2]

The booming farm economy of the 1970s introduced private-sector lenders to increasing competition from government-sponsored lending institutions. The Farm Credit System was created in 1933 as part of a series of New Deal initiatives aimed at providing financial security to farm debtors. As a federal program, it provided start-up money to the Federal Land Bank and the Production Credit Association (PCA), rural lending agencies that became borrower-owned in 1947 and 1968, respectively. To obtain credit from either agency, farmers purchase stock worth between 5 and 10 percent of their loan's value. Regardless of loan size, each borrower is entitled to vote on institutional issues of governance and policy. The Land Bank was authorized to issue real estate loans, while the Production Credit Association supplied short-term loans for chattel and operating expenses. In 1987 the Land Bank and PCA were merged to form Farm Credit Services.

The congressional mandate of the Land Bank and the Production Credit Association has been to help qualified farmers who have difficulty obtaining credit from a private lender. When these agencies repaid their federal loans, they technically became private institutions, and from the early 1970s to the mid-1980s, they rapidly increased their share of farm debt relative to other lenders.[3] Economic conditions appeared to warrant this more aggressive posture, and many farmers welcomed the opportunity to try their hand at farming on a larger scale. Since Farm Credit agencies relied on adjustable rate loans, loan officers could afford to offer competitive interest rates and take risks that private lenders were unwilling to take—often drawing disgruntled borrowers away from community banks, where the attitude could seem condescending and, in an era of quick profits, more than a little stodgy.

The Farmers Home Administration (FmHA) was created in 1947, the postwar successor to agricultural credit programs created by the government during the depression.[4] Until 1994 FmHA was the agency with the U.S. Department of Agriculture that provided federally subsidized credit and technical assistance to rural America.[5] From its inception, the primary purpose of Farmers Home has been to provide low-cost financing to farmers who are denied credit by other lenders—thus its moniker, the "lender of last resort." To offset the riskiness of its loans, FmHA has traditionally combined financial assistance with close supervision and locally sponsored education programs. The lending relationship is frequently characterized as a form of apprenticeship in which county supervisors take promising young farmers under their wing and guide them through the ins and outs of farming until they "graduate" by refinancing their loan in the private sector.

Where farmers enter the credit structure—at the private, quasi-private, or public level—determines far more than how much money they can borrow or what it will cost. For the farm economy is structured by values and expectations that are as much cultural as they are economic. Who is able to borrow from whom, what reciprocal obligations are entailed, and how the repayment of debt is negotiated in the event of hardship are matters of custom as well as political conflict. Traditional wisdom frequently clashes with contemporary sensibilities when it comes to questions of who deserves credit and why. Nowhere is this dilemma more apparent than in the conflicting criteria private- and public-sector lenders use to determine credit-worthiness. Where individual character and family reputation come first in private forms of credit, individual merit and personal achievement are the primary concern of public-sector creditors.

Cultural Credit

When the blizzard of 1940 swept across the state of Minnesota, the Van Hoffs' five hundred sheep were caught between the pasture and the barn.[6] Buried beneath snow eight to ten feet deep, the sheep would have suffocated were it not for the efforts of the entire family—six children and two adults. After five days of walking across the surface of the snow, poking rods into the ground, and digging wherever steam came up, they were able to rescue the entire herd. The drama of the event would have stayed with eight-year-old Chester regardless, but in his mind it is fatefully linked to his father's illness and death later that winter. The youngest of the Van Hoff children, Chester was determined to begin farming as soon as he and his older brother, Joseph, were able to acquire credit on their own. He recalls what it was like to apply for their first loan in 1948:

We each had a cow that we started with. That was our total equity. We went in to the banker to ask them if we could buy four heifers. At that time, the banker—he was about seventy-five to eighty years old, and he was a short little fellow. He'd stand on one side of the counter and you'd stand on the other side, and he'd tip his glasses down and look at you over his glasses. Just study you, you know. And he stood there for the longest time; and we were standing, just sweating, wondering what was going to happen. Red and flushed and nervous. And he finally agreed to let us borrow money for four heifers. When we went back into the bank to get the money, we went through the same thing again. We stood on one side and he's looking at us, studying us over. He didn't say yes, he didn't say no. Pretty soon he says, "OK—but you're going to have to pull an awful lot of tits to pay for those!" I think he was wondering if we could handle farming, because we were pretty young. Joseph was twenty-one—but I was seventeen and legally couldn't borrow money.

Looking back on that day of judgment, Chester marvels that, despite his youthful nervousness, getting started in farming back then had been so "easy." Not only had it been possible to start a dairy operation with so few cows, but they had been able to do it on the basis of so little equity. Chester knows that this first loan was a test—based as much on the banker's assessment of their character as it was on their collateral. Indeed, that is the charm of his story. Between the lines swells the pride of two fatherless boys who find that their family's reputation redounds well upon them. For in essence, the economic opportunity they were given depended on the "credit" they received for a moral character they had not yet demonstrated.

When farmers like Chester Van Hoff suggest they were issued loans on the basis of their character, they do not exaggerate. Among established farmers in Star Prairie, credit based on the reputation or "family name" of the recipient is the preferred way of doing business. While economic considerations play a role, they often take a backseat to the value of the social obligations that debt will create or reinforce. Whether the sale of farm property is within the family, between neighbors, or financed by the local bank, the collateral securing these loans is as much social as it is economic. It is membership in the community, not economic capital per se, that qualifies a person for credit based on what sociologist Pierre Bourdieu has called "social capital."[7] For the benefits of group membership must be *earned,* not bought, and the "investment in sociability" can be considerable—spanning a lifetime, if not generations.

The Glacial Lakes Bank was founded in 1894 by Zachary Stow, a German immigrant who made his way to western Minnesota and set up business in the lobby of the village hotel forty-eight years after the territory was made a state. One hundred years later, I am greeted warmly by his grandson, Zachary Jr., the current president of the bank. Silver hair gleams on his temples and his smile is catching as he presses a self-published centennial history of his bank into my hands. We are no longer in a Victorian hotel, but pleasantly soothed, on a muggy July day, by the air-conditioning that filters through the glass-enclosed offices of a 1970s ranch-style building. Ground breaking for the new bank took place in 1980, at a time when plans a decade in the making could still go forward without a hitch.

Looking back on his forty years of experience as a bank president, Zach is candid about the criteria he has used to authorize loans.

ZACH: Years ago you could drive through the country and you could look at a set of farm buildings and you could tell how they were kept up. That told you the story of what the farmer's thinking was. You know, I used to have a policy that I think worked pretty darned good. If somebody asked me for a loan and I didn't know him before, I'd say, "We'll have to think it over for a day, could you come in tomorrow?" And I'd walk back out [of the bank] with him, because I wanted to look at the inside of his car. Because I thought the inside of his car told me the same story that was upstairs [*pointing to his head*]—the same condition was up here that was in there. If it was a *mess* in there, then I thought it was a mess up here too [*laughing*]. It might be dumb, but it worked pretty good!

KATE: I imagine the same would be true if you drove past the farm and took a look.

ZACH: Exactly. If there's a big mess, machinery all over the place.

> No paint, I can understand that. But if it's kinda neat around there—it's just one of my little quirks, I guess.

Banking rules and regulations have changed over the years, and Zach is no longer able to exercise as much personal discretion in making farm loans. Yet he remains adamant that his "policy" was as good, if not better, than the mountains of paperwork that now replace it. For no form in the world can tell you what a person's moral character is—and that, for independent bankers of his generation, is the key to a successful loan.

Clayton Miller, retired vice president of the Traverse Junction Bank, heartily agrees. As elsewhere in America, a large percentage of the loans a bank makes are to young people at the start of their working lives. Being a good judge of character is absolutely critical, Clayton argues, for in the end, a person's "work ethic" is a banker's only guarantee of repayment:

> The younger generation—and to a certain extent we all go through that phase in our lifetime—when we're young, we're not worried about repaying our debt. We're young, we're energetic, we've got the world in front of us, and all we've got to do is borrow the money and get to work. Which of course is basically true—if they want to work and will work, you know? There are so many things to consider in lending. The way the people spend their money, it's a great concern of ours. If you see a man in the liquor store every day, playing cards and drinking beer, rather than being out on the farm and working, those are things that have a big bearing on whether or not he's going to be able to get credit.

Ultimately, what the recipients of bank loans get credit *for* is having the cultural disposition to be good farmers. Being a member of a farming community, growing up on a farm, and, increasingly, acquiring an advanced degree in crop science, animal husbandry, or agricultural economics are all indicators of what is, in the final analysis, an integral part of the person.[8] It is the inner capacity to farm—the habits of mind and body demanded by this kind of work—that bankers look for in a loan applicant. Zach Stow's "neatness policy" or Clayton Miller's "liquor store check" are creative strategies for ascertaining what they—and their board of directors—most want to know: Do these individuals have "what it takes" to do what they say they want to do? Simply having the money to buy a $100,000 tractor does not enable a person to do anything productive with it. Farmers add a crucial "know-how" to the equation, and this is cultural competence they receive "credit" for when they pass muster at the local bank.

To a greater or lesser extent, all forms of agricultural credit rely on an assessment of the applicant's ability to farm. But only at locally owned banks is the "recognition" of cultural capital considered a legitimate business practice. In this respect, community banks have much in common with voluntary association loans, such as those issued by the Lutheran Brotherhood, and the loan contracts negotiated with family members or neighbors. Originating in the depression when few banks would take on the risks of a real estate loan, "contract-for-deed" agreements have long been an important mode of farm transfer in Star Prairie. When farmers sell their land on contract-for-deed, they "lend" their farm directly to the buyer on a repayment plan both have agreed upon. When the contract is fulfilled, the deed is handed over to the buyer. These private arrangements often mimic the effects of an intergenerational transfer insofar as they provide the senior generation with a steady interest income and the option to use a "balloon payment" at the end of the contract period to make bequests to nonfarming heirs.

Private lenders in Star Prairie make no apologies for the fact that their loans are often based on ties of kith and kin, not hard-and-fast economic rules. Matt Rundgren, a young loan officer at the Star Prairie Bank, recites a formula he has come to know well:

If you want to come in and start farming, what is our thought process? There are four "Cs" that we look at. They're characterized as capital, character, capacity, and collateral. As far as *capital* goes, what does your balance sheet look like? How many assets do you have? How many liabilities? What's your net worth? What do you have to put on the line? Do you have cash readily available? How liquid are you? *Character.* Does the person have the background? Was he raised on a farm? Is he coming out of the Cities? Is he bright? Can he pick this stuff up? Does he pay his bills? How has he handled his past credit? Savings, checking accounts, how disciplined has he been? So you make a judgment, is the guy gonna succeed? Does he have what it takes to manage that successfully? *Capacity.* Does the unit—when you put together the cash flow—have the income, the expenses, as they're projected, to meet the projected repayment? How's the debt gonna be structured? The interest rates, the annual payments compared to the annual income, taking out the living expenses, how much is that? You want to mortgage for adversity. You want to try and realistically come up with a plan. *Collateral.* Where are we at collateral-wise? What are they offering? If they're buying a farm, how much are they putting down on it? What is the condition of the collateral, if it's machinery? Looking at it, all of those are slices in the decision.

Matt smiles and adds, "There's actually a fifth 'C,' but it's not easy to explain." He taps his pencil on the desk, seeming to hesitate. "It's basically *common sense,*" he says. "You don't look at it and think, Oh, we're well secured; we'll make the loan. If the guy don't have the capacity to succeed in that unit and to manage it, it's probably going to end badly." Whether it's called common sense or "a little quirk," finding informal ways to evaluate a person's cultural capacity to farm remains central to private forms of credit. To newcomers in the community, local lending relationships can appear cliquish and unfair. But bankers are not the only lenders in town.

Individual Merit

Dick and Diane Porter moved to Star Prairie as newlyweds in 1970. Natives of southeastern Minnesota, they bought 180 acres of the only land they could afford—the dry sandy land in Bonanza Valley. Through a combination of off-farm employment and diligent efforts with corn and soybeans, they were able to capitalize on the high prices of the 1970s and plow their profits back into the purchase of an additional 240 acres. Irrigation helped get them through the drought of 1976, and by the late 1970s, they felt they were ready to begin renting more land and building the large-scale potato operation they aspired to. But they found local bankers unreceptive. To this day, Dick rankles at how cavalierly his loan requests were denied:

[The banker's] answer was—he belittled me. He said, "Well, gee, you haven't really gained a whole lot here in about three years." My answer to him could have been, "What's *your* net worth? And what's *my* net worth?" Granted, mine's on paper and all that, but if I had a sale tomorrow—I'm worth something! And yet you go to the bank and they say—you're worthless; you haven't got anything. One of the things I've realized is that when you go into a bank, it's almost like—an individual. Whether they don't like your hair or your shoes, or they just don't feel comfortable with you, or the friendship factor or whatever it was—if they *choose* to not give you a loan, that's the end of it.

Dick feels that the value of his accumulated assets made him a good credit risk and worthy of the bank's support. By the end of the 1970s, with land values rising throughout the county and the capital improvement made by their irrigation system, the Porters had, in fact, made steady gains in their "net worth"—the total value of their assets minus their total liabilities. In purely economic terms, the market value of their assets put their net worth at close to a quarter of a million dollars. With this strong equity position, Dick insists, he deserved better treatment at the bank—no matter who his friends were or what he looked like. Feeling insulted and "belittled," he took his business elsewhere: to a branch office of the Farm Credit System.

Farm Credit borrowers share with farmers who receive credit in the public sector a visceral mistrust of how credit decisions are made by the "banker clique." To their mind, there is something "undemocratic" about a system that confers economic advantage on the offspring of farm families who have lived in the community for generations. What kind of society is this, they demand, if you can't borrow money from the bank unless "you've been here 150 years" or have "at least one generation in the cemetery"? Who your parents are, how long you've lived in the area, or what the inside of your car looks like are not, in their view, legitimate lending criteria. To farmers who feel like victims of "discrimination" in the private sector—often complaining of "shiny knees" or "wearing out their pants" trying to get a loan at the bank—the chance to be judged on their individual merits is considered a fundamental right and a critical affirmation of their self-worth.

To the extent that both the Farm Credit System and the Farmers Home Administration are legacies of the New Deal, they are expressions of a moral vision that differs in key respects from that which governs private lending arrangements in Star Prairie. Historian Catherine McNicol Stock has written with insight of the cultural conflict triggered by New Deal programs on the Great Plains in the aftermath of the depression. During these years, the old middle class of shopkeepers, farmers, and independent professionals was forced to make room for—and cede authority to—a "new middle class" of experts, advertisers, agents, and advisers. In this encounter, the old middle class lost exclusive control over local decision making, as rural communities were increasingly exposed to the "people, policies, and ideals of postproducerist America." Representatives of this emergent national culture "saw America not as a world of producers but as a society of consumers ready to receive their services, ready to judge others more on the basis of personality than character."[9]

Vestiges of this depression-era conflict continue to linger, I believe, in the competing visions of moral worth at play in the contemporary debate between public- and private-sector borrowers. At issue is the underlying question of who is truly authorized to decide the credit-worthiness of farmers: the local community or the national government? Farm Credit borrowers occupy a precarious middle ground in this debate, straddling a fine line between public and private forms of credit. Although they are stockholders in their own lending institution, they are still subject to federal directives and the "top-down" management style of multibillion-dollar corporations. There is frequent turnover of loan officers, and bureaucratic anonymity plagues Farm Credit agencies as much as it does insurance companies, on the one side, and Farmers Home, on the other. During the farm crisis, Farm Credit borrowers often found themselves in a cultural "no-man's-land" between the community's desire to rally around its own and its refusal to take on problems not of its own making.

In the years following World War II, supervisors hired by the Farmers Home Administration were often drawn from their home counties and remained in their positions until retirement. This was true of Skip Gorder, who, as an FmHA supervisor, put his personal stamp on the character of farming in Star Prairie. Gorder was well-known for his conviction that the best way to begin farming from scratch was to milk cows. To an ambitious farmer like Melvin Johnsrud, however, such wisdom seemed tethered to a bygone era. In the early 1970s, crop prices were hitting all-time highs, and it made no sense to him to run valuable grain through dairy cows:

The biggest mistake I made [when buying this farm] was buying milk cows. You cannot afford to feed $5 wheat or $3 corn to milk cows when milk is only worth $4 a hundred[weight]. In order to get money to purchase the farm, the FHA guy here in Star Prairie demanded that you milk cows. That was the only way they would approve a loan. I bought the farm in May of '72 and then, when the Russians bought all the grain, the market prices went way up. There I was feeding all that expensive grain [to dairy cows]. I had sixty bushel [an acre] wheat that year that was worth $5 a bushel.

I paid $200 an acre for the farm. If I'd have had the whole farm in wheat and never saw a dairy cow, I'd have been rich. But everybody knew Skip Gorder's history: you had to milk cows in order to get a loan from FHA. It was his philosophy of life.

Generally speaking, Skip Gorder's preference for milking was not a bad "philosophy of life"—and one the older generation still subscribes to. Because a dairy operation generates a monthly income, it is possible to monitor the "cash flow"—the farmer's income and expenses—on a regular basis and detect management problems as they arise. It also requires daily labor on a fixed schedule, leading some lenders to believe that it "builds character" in young farmers prone to indulge in less industrious pursuits. For every farmer like Melvin who resented the constraints of FmHA borrowing, there were others like Ray and Alice McClelland who welcomed the opportunity.

The McClellands began looking for land in western Minnesota in 1969. Having grown up on a dairy farm, Ray liked to milk, and he found a kindred spirit in Star Prairie's FmHA supervisor. Although the farm they purchased had previously supported beef cattle, Ray remembers, they felt it was ideal for dairy and Skip Gorder agreed:

I told Skip I want to put in a dairy because that's what I know the best. And he liked dairy farming. That was what a lot of Star Prairie was at the time. But most of the young guys would say, "Geez, I don't want to milk." But it was the only thing I knew. This farm was [meant to be] a dairy farm. It had a lot of low land, and you could raise a lot of feed on it. Gorder went along with it and we built a barn, a real conservative barn, and we remodeled a shed. So by the time I got in here and got settled down, after a couple years, I was in debt about $100,000.

In the scheme of things, being $100,000 in debt at the beginning of the 1970s was not out of the ordinary, especially for young farmers with no equity of their own. Within a few years, with the dairy operation running smoothly and land values on the rise, the McClellands were well on their way to achieving a modicum of financial stability. Had it not been for a tragic accident in 1979, they might have been spared the worst of the farm crisis. As it was, however, they lost their entire dairy herd to an electrical fire that tore through the barn late one night. Walking through the charred ruins was one of the worst days of their lives, Alice and Ray recall—but it was only the start of their troubles. Although the McClellands received insurance money, it was not enough to cover the cost of rebuilding their dairy. Hoping to use this money as the foundation for a new loan, they looked to Farmers Home for refinancing. But by the end of 1979, the federal agency was under increasing political pressure to clear out old accounts, "graduate" eligible borrowers, and respond to the new policy directives of the incoming Reagan administration. Internal reorganization was ushering in a new generation of county supervisors—young college-educated government employees who were eager to prove their mettle and run their county office "by the book."[10]

ALICE: We [had] changed FHA man; old Skip retired. When Skip said something, you could believe him and trust him and he was for you. So we changed; a young FHA man came in.

RAY: Young college boy.

ALICE: I'm sure he was told to "weed out farmers," and that's exactly what he did.

RAY: He lied to ya.

ALICE: To everyone. You couldn't believe what he said. He walked out here after the barn burned and his question was, "Well, when are you going to sell out?"

RAY: "You're done."

ALICE: Skip Gorder still lived in town, and we went and talked to him. He couldn't believe it.

RAY: He said, "Get it refinanced. Get out of there."

ALICE: The young FHA man, every time you'd go to do something, he was always bucking you, always fighting you. There was never—he was never trying to really help you.

RAY: That guy ruined this county, as far as young farmers.

KATE: Was it just that individual or a whole change of philosophy?
RAY: There was a whole change of philosophy.

Ray and Alice were given an ultimatum that left no room for negotiation: if they did not pay off their loan at once and in full, Farmers Home would initiate foreclosure. As it happened, land values in Star Prairie were approaching their peak, and the McClellands had accumulated substantial equity in their 460-acre farm. Using the insurance money to repay FmHA and their land as collateral, they qualified for a mortgage from Travelers Insurance Company. Budgeting on a shoestring, they managed to rebuild their barn and began raising young dairy cows. At any other time in the postwar period, this "graduation" from public- to private-sector financing would have been a step up in the world. But the Federal Reserve Board's newly restrictive monetary policy had begun to take effect with a vengeance. Instead of the subsidized interest rate they had been paying at Farmers Home, Travelers' adjustable-rate mortgage was 12.5 percent and going higher. After several years of low livestock prices and double-digit interest rates, their debt finally overwhelmed them—and in the fall of 1984, Travelers issued a notice of foreclosure.

As the "credit crunch" began to take hold, farm families throughout the United States began to reexamine their "debt-to-asset" ratios—their amount of outstanding debt divided by their total assets. In January 1984 the Federal Reserve Board issued a report estimating that 8 percent of the nation's farmers had debt-to-asset ratios of over 70 percent, and that 11 percent had ratios between 41 and 70 percent.[11] Together, these two groups—19 percent of all farmers—held 63 percent of the total farm debt. These highly indebted farmers were at the greatest risk of financial failure and were, on the average, forty-seven years old.[12] As useful as these statistics are, they offer little insight into how farmers with different socioeconomic backgrounds responded to the farm crisis. To understand what debt and dispossession means to those caught up in this crisis, we must attend to the social norms and obligations that constitute the community's "moral economy."[13] For it is here, in the institutionalized measures of moral worth that structure the agricultural credit system, that farm loss acquires its cultural meaning. Whether farmers borrow from a lender in the private or public sector—and whether they were granted loans on the basis of cultural credit or individual merit—had a much greater impact on their experience than the overall amount of their debt. What mattered when times got tough was not how much money you owed, but whom you borrowed it from.

Some of the owner men were kind because they hated what they had to do, and some of them were angry because they hated to be cruel, and some of them were cold because they had long ago found that one could not be an owner unless one were cold. And all of them were caught in something larger than themselves. Some of them hated the mathematics that drove them, and some were afraid, and some worshipped the mathematics because it provided a refuge from thought and from feeling. If a bank or a finance company owned the land, the owner man said, The Bank—or the Company—needs—wants—insists—must have—as though the Bank or the Company were a monster, with thought and feeling, which had ensnared them.

—John Steinbeck, *The Grapes of Wrath*

4 Primal Scenes

Early one morning in the summer of 1933, Farmers Bank was robbed by two masked men who, the story goes, were dressed as farmers. The cashier, bookkeeper, and customers were forced to lie on the floor while the bandits scooped money from the till. When ordered to open the safe, the cashier explained that it was protected by a "time lock" security system and could not be opened until later in the day. Meanwhile, a woman approached the bank and saw what was happening. She ran across the street to the café and alerted the proprietor. He snatched up his shotgun and opened fire on the men as they ran out of the bank. The man with the money was killed; the other escaped and was captured days later.

Not until chatting with Phyllis, a teller at the Star Prairie Bank, did I learn that the bank where I have been cashing checks all summer was, in fact, the present incarnation of the Farmers Bank, which had built a new bank on the site of the old café in the 1950s. Pointing across the street at the old Farmers Bank building—now the cooperative feed and fertilizer store—Phyllis retells the story of the robbery as if it took place yesterday. "That's where it happened," she concludes, enjoying my evident surprise. Perhaps it is always eerie to witness the spot where someone has died, but there is something more, something else, in Phyllis's tone of vindication. It is her pointing finger and the niggling detail that the culprits were "dressed as farmers."

If a community can be said to have a "primal scene"—an episode in which the "truth" of the underlying social order is witnessed and then repressed—the drama of a thwarted bank heist would qualify as such a scene in Star Prairie. For central to a capitalist economy is a latent antagonism between those who make a profit by lending their money and those who must sacrifice some part of their labor in order to borrow it. In a community based on the social relations of market exchange, a cultural premium is placed on *suppressing* awareness of the inherent inequality of these economic positions. In fact, the "appropriate" response of those who witness the primal scene consists in precisely this. Like the vigilant woman and heroic café owner in the Farmers Bank robbery, maintenance of social order depends on punishing those who transgress community norms by "acting out" the conflict we all know lies beneath the surface of everyday life.

Villains and Victims

When Rudy Voslek decided to help his two sons start farming, the local bank offered its immediate support. Rudy's grandparents were early settlers in Star Prairie, and the Voslek family had worked with the same bank for two generations. Hoping to build a state-of-the-art hog barn and bring his sons into the operation as full partners, Rudy sat down with his banker and worked out a financial plan agreeable to everyone. The bank authorized a loan for a new barn, livestock, and machinery, while the Federal Land Bank approved a loan on additional land. For much of the late 1970s and early 1980s, everything ran according to plan, even as rising interest rates ate away at an already slim profit margin. During this period, Rudy's wife was diagnosed with cancer. Her treatment involved frequent trips to the Mayo Clinic in Rochester, Minnesota, and the Vosleks' insurance did not cover many of the medical costs. His savings depleted, Rudy was caught in a bind when land values crashed and the Land Bank demanded more collateral to secure its loan.

There appeared to be only one way to rectify the situation. Rudy's mother and sister had each inherited a portion of his grandparents' estate, and these parcels of land were unmortgaged. The bank was willing to secure the Vosleks' loan with the Land Bank if they put up this land as additional collateral. Rudy's mother was willing; his sister was not—she asked to be paid in full for the land before she signed it over. The involvement of extended families in a farm operation is not unusual, though as the actions of Rudy's sister make clear, there are limits to the risks that kin feel obligated to take.[1] After a tense meeting with the banker, lawyers, and family members around the kitchen table, the deal went through. In retrospect, Rudy says with a sigh, "And my sister was the only smart one in the bunch." Less than three years later, the bank abruptly canceled the Vosleks' line of credit, making it impossible for them to pay for routine operating expenses. Federal Land Bank moved to foreclose, and every scrap of collateral property—save for the house that Rudy lives in—was lost.

As Rudy gets to this point in his story, we are joined by Tessa, his second wife. An old family friend who helped Rudy through the death of his wife and the loss of the farm, she now sets aside the potatoes she has been peeling and sits down at the table with us. Rudy takes her hand and says that he "learned a lot" from his experience. I ask him what he learned.

RUDY: Not to trust. I think we had full trust in our banker in the earlier years. We relied on him. I don't think we did anything where I wouldn't talk to my banker first. My dad did the same thing, and of course I had done the same thing too.

TESSA: And you were working with the same bank.

RUDY: I'm not anymore!

TESSA: It's like when you see these old movies where the banker goes out to the ranch, and the husband has to hide, and the wife is trying to keep the mortgage going on this place. Even back then, the banks were looking out only for themselves—*not* the person that they borrowed the money to.

RUDY: Typical situation.

TESSA: Yeah, so typical! It has just gone that way through all the years.

It is telling that the Vosleks recognize their situation in an old movie Western of villains and victims.[2] At the core of this script lies the primal scene of the Farmers Bank robbery with the key roles reversed: here it is the family "trying to keep the mortgage going" who is victimized by the banker's actions. No one rides in to rescue them, however, and in the end, it is the banker who has his way. As witnesses to this primal scene, Rudy and Tessa—having learned "not to trust"—no longer suppress awareness of it. Nonetheless, a certain forgetfulness has fallen over the experience. Like an old movie, the outcome of the underlying conflict is already known—as "it has just gone that way through all the years."

At the Star Prairie Bank, I am assured by vice president Frank Tostrud that there is no economic conflict of interest between farmers and bankers.

FRANK: Our best interest is the farmer's best interest. There's no conflict there.

KATE: And there never would be?

FRANK: No. They may *perceive* that there's a conflict there, but there really isn't.

KATE: Explain to me how that might be.

FRANK: Well, if the farming unit can produce only enough feed to handle thirty milk cows and the margin of profit in producing milk is such that you've got to have sixty milk cows, then you either have to increase the size of the unit or you have to quit. If there isn't any equity capital there, you have no way to increase the size of the unit—because that also spells absolute failure.

> You can't operate a million-dollar business with three dollars of capital: you're going to fail. Another person in that same business with $350,000 of capital is gonna succeed—just because of the cash flow. So there is no difference between what's best for the farmer and what's best for the banker. It's the same. But you have to work together to find it.

I ask Frank if adverse economic conditions—such as the rising interest rates, falling land values, and low commodity prices of the 1980s—might introduce a conflict of interests. Wouldn't the bank be interested in recovering its investment and the farmer in continuing to farm? Smiling, he replies:

> To continue farming at a loss? You see, Kate, you said that the interest rates had risen and the values had plummeted and it's to his interest to keep farming. You just described a situation that puts him in a position of operating at a loss. How is operating at a loss in his best interest to continue?

At first glance, Frank's argument seems indisputable—and, at the level of public discourse at least, one that farmers no doubt agree with. For who would want to operate a failing farm? But much is left out of this official story. In particular, focusing exclusively on whether a farm should continue in operation begs the question of what happened to the farms that failed. If it is not in the farmer's interest to *run* a nonproductive farm, then what is the value of that farm to the bank? Unless the lender is able to recover the original loan amount through the timely resale of collateral property, repossession does not always make financial sense. Land, buildings, machinery, and equipment can produce no income for the bank laying fallow or sitting in a warehouse. During the farm crisis, when farm incomes were low and interest rates were rising, there was no guarantee of finding a ready buyer. Yet, in conversations with me, no lender acknowledged that this state of affairs had any bearing on how their loan contracts were enforced. Implied, if not stated, is the belief that economic rationality—indeed, the well-being of the economy itself—depended on finding ways to execute maximum loan collection. Sometimes that involved restructuring a borrower's debt, either by lowering payments and extending the payment period or by "charging off," as a loss to the bank, some portion of the original principal or interest. More often, however, lenders simply moved to collect what they could, as fast as they could—eager not to be left holding the bag in a dead resale market.

Talking about loans made by the Glacial Lakes Bank in the 1970s, vice president Harvey Beckett draws an analogy to his father's decision to spray his fields with DDT in the 1950s. Once the dangers of this pesticide were known, farmers no longer used it. Does this mean that his father's decision was wrong? Of course not, Harvey says. The decision was good at the time, even though it would not be made today. Likewise, he argues, every loan is "good" at the time it is made:

> You always make good loans; they just turn bad. If I sit down across from you today and I'm going to make a new loan for a new applicant, I would believe that that's a good loan. There might be some circumstances that change within the next five or six years—either in regards to interest rates or maybe it's decisions that you or I have made. That loan might go bad and I might have to take a charge-off on it. But I don't sit down today with the thought that I'm going to charge off part of this loan in six years, or I'd never make it. I've heard that statement in banking many times: *You never make a bad loan—they just turn out that way sometimes.* Some of the time it's totally beyond any of our controls. And a lot of the time in the early eighties it *was* beyond any of our controls. It just happened.

Like Frank Tostrud, Harvey Beckett is concerned with putting the position of community banks during the farm crisis into a moral perspective. Who could have known at the time "nonperforming" loans were made what the future would hold? Once again, it is hard to imagine that farmers would disagree with this public appeal to human fallibility. Yet here, too, much is missing. There is no talk of how loan officers had something to gain from encouraging farmers to take on as much debt as they were legally entitled to. Nor is there open recognition of the fact that although both parties "signed on the dotted line," only one party—the farmer—is held personally accountable for repaying the loan that "looked good" to both at the time it was made. Instead, bankers are quick to point out that the losses they were forced to take—either through foreclosure or loan restructuring—were ultimately losses to the community as a whole. As Harvey contends, defaulting farmers posed a threat to "all of us," not just the bank:

I know the perception out there. The bank took a loss? Hey, big deal, they got a lot of money. But if they follow this thing through—it's no different than shoplifting, OK? How does an owner pay for shoplifting? He builds that into his margins. So he charges more for his product in order to cover the shoplifting. When the Farm Credit System took the hit, they went to the federal government and they got money from them. Now, who's that? That's all of us, as far as taxpayers. So all of us chipped in. The same thing happened with the bank and the farmer and the community. The community is the one who took the hit. They absorbed part of it along with the banker. Those who have deposits here or are still borrowing money here. Let's say they're retired people with a CD at that time. Maybe instead of getting 9 percent on their CD maybe we only paid them 8. Maybe the person who was borrowing money at that time, instead of what would have been a going rate of 12 percent, maybe they would end up paying 13. So both the depositors and the other loan customers lost, because we had to make up those shortages someplace else.

The notion that distressed farmers were "stealing" from the bank would seem to suggest a conflict of interests. Yet Star Prairie's bankers do not see it that way. Rather, as Harvey believes, they are simply acting on an obligation to protect their customers and shareholders:

You've got to imagine who we work for. I guess you'd say I work for the board of directors or the owner. In some degree, [we're like] the FDIC [Federal Deposit Insurance Corporation]. They're out there to protect the depositors, and that's who we're working for too. So you might say, I'm out there fighting against this farmer to protect the rest of the community. It's not me against them. I'm an intermediary trying to protect the depositors here. So it was kind of a neighbor fighting against neighbor type of thing. But I'm sitting in the middle and I'm the one that gets the blame.

Perhaps more so than other lenders, community bankers are aware of the need to justify their actions in terms that make cultural sense to the majority of their customers. As the legitimacy of the foreclosure process came under sustained attack from the grassroots farm movement, bankers were at pains to distinguish their "honorable" way of doing business from that of lenders at the government-sponsored and government-administered levels of the credit structure.

Ethical Alibis

In the late 1980s, Erik Ekholm left Farm Credit Services to become a vice president at the Star Prairie Bank. Happier in his new position, he feels that community banks were able to respond to farmers in crisis with greater compassion than Farm Credit did. Recalling his years as a Production Credit loan officer, he speaks of himself as a lowly "grunt" in the trenches of a vast bureaucracy—forced to carry out orders that seemed needlessly "cruel":

[This administrator and I] were having an "I told you so" session. Most of us grunts were still saying, "Hey, we've got to do something here to work some of this stuff out." This guy said, "We're not going to do any of that until the farmer is living in a trailer house on the sand flats!" Basically, we're going to *bulldoze* him into the ground. It was the cold, hard, cruel way of doing business at that time. If you have three aces, why give up anything, right? If the way in which you get maximum collection on a loan is to liquidate all of the collateral, why would you *not* do that? Well, one of the reasons why you wouldn't do it is that you ruin the value of other people's collateral. You know, Farm Credit has thirty to forty thousand acres of land, every small bank in the area has a bunch of land, and you've sold all this machinery at fire-sale prices. Well, you just wreak havoc with the values of machinery and real estate. Not to say [values] shouldn't have come down, but should they have gone down as low as they went? No, they shouldn't have.

Erik is not alone in suggesting that the farm economy's deflationary spiral was due in large part to the draconian repossession tactics of the national Farm Credit System. The Production Credit Association and the Federal Land Bank were, by the mid-1980s, the lenders with the largest share of agricultural debt, much of it incurred in the 1970s. During the peak years of the crisis—1984 through 1987—these agencies experienced significant losses in dollar amounts. In comparison, however, banks accumulated as much or more bad debt than Farm Credit, particularly when viewed as a percentage of all agricultural loans (see figures 4 and 5).[3] Thus, although local patterns of foreclosure might have created the impression that Farm Credit was responsible for flooding the resale market and driving down the value of other lenders' collateral, the national picture suggests equal culpability.

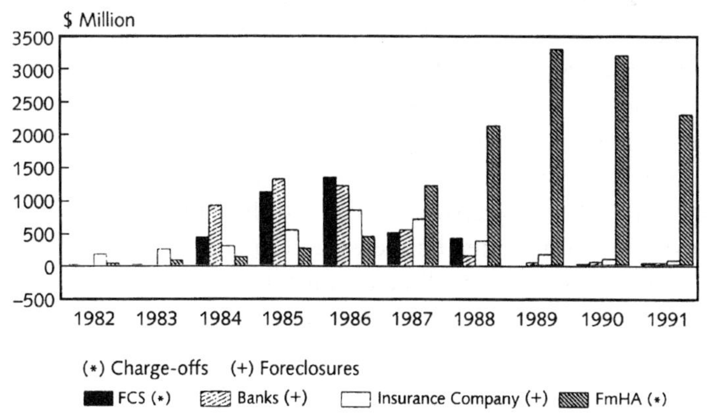

Figure 4 Dollar Amounts of Lender Losses, 1982–1991
Source: USDA Economic Research Service, and Kenneth L. Peoples, David Freshwater, Gregory D. Hanson, Paul T. Prentice, and Eric P. Thor, *Anatomy of an Agricultural Credit Crisis: Farm Debt in the 1980s* (Lanham, Md.: Rowman & Littlefield, 1992), fig. 2.3, p. 38.

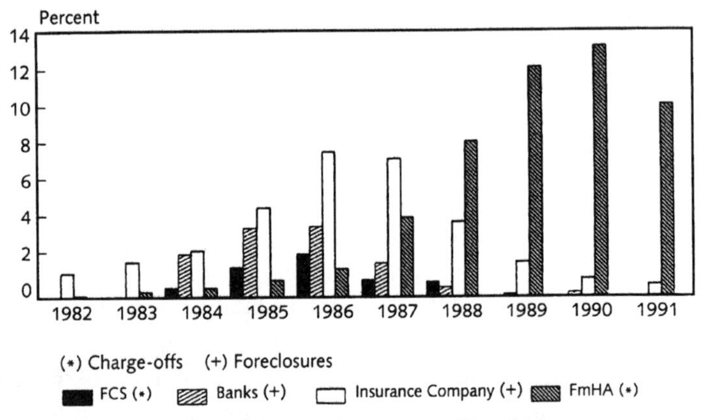

Figure 5 Lender Loss Experience as Percent of Total Loans, 1982–1991

Source: USDA Economic Research Service, and Kenneth L. Peoples, David Freshwater, Gregory D. Hanson, Paul T. Prentice, and Eric P. Thor, *Anatomy of an Agricultural Credit Crisis: Farm Debt in the 1980s* (Lanham, Md.: Rowman & Littlefield, 1992), fig. 2.4, p. 38.

But local perceptions of lender actions during the crisis were also shaped by the cultural assumptions that distinguish different kinds of debt. Of particular interest in this regard is Erik Ekholm's claim that he had no choice but to act as he did when he was employed by the Production Credit Association. To be a "grunt" in a bureaucratic system is to have no say in which wars are fought or what tactics are used. Anthropologist Michael Herzfeld observes that bureaucrats and clients alike rely on the "ethical alibi" of a "heartless system" to explain why people caught up in large organizations are compelled to act in impersonal ways.[4] In the official story of the farm crisis, community bankers often emerge as local heroes—or at the very least, smelling cleaner than employees of Farm Credit Services, insurance companies, or the Farmers Home Administration.

"When the financial people want to get out their money," says Chester Van Hoff, "they were your best friend—but when it turned the other way, why, they were a completely different person." Had it not been for a "sixth sense" about what was happening in the 1980s, Chester and his wife, Ivy, could easily have shared Rudy Voslek's fate. In 1984, when their oldest son, Philip, married and wanted to start farming on his own, they purchased a farmhouse and building site for the young couple. Using this property to secure loans with Farm Credit's Production Credit Association and Land Bank, Philip and his wife were able to buy a new tractor and eighty acres of land. At one point in the negotiations, Chester and Ivy were encouraged to mortgage their own home to "back" what could have been an even larger loan. They demurred, feeling guilty at the time that they were not doing more to help their son. But less than a year later—before the first loan payments were even due—Farm Credit initiated foreclosure and swiftly moved to repossess the newlyweds' land, house, and building site.

I am flabbergasted. "How could they do such a thing?" I ask.

Chester shrugs. "Well, they just called it a *bad loan* because Philip owed more than the property was worth." Within the year after the loan had been made, land values in Star Prairie had dropped almost 30 percent, and the Federal Land Bank was determined to cover this loss with the sale of collateral property. Before taking final action, Farm Credit approached Chester and Ivy, asking them to make up the shortfall and resecure the loan, but they flatly refused. In their view, their son had done nothing wrong—absolutely nothing to deserve this outrageous treatment. Philip's foreclosure occurred before Minnesota's mandatory mediation law went into effect, so the elder Van Hoffs never had the opportunity to give his loan officers a piece of their mind.[5] But it is not hard to gather what they might have said.

CHESTER: The financial people were just out to grab. They were in trouble and had a lot of bad loans and they were taking any property they could get their hands on and reselling it and getting the cash.

IVY: As far as Federal Land Bank, Production Credit, and a lot of these loan establishments, they were looking for money from the government at that time and they wanted the government to be behind them. So they were *making* everything look bad so that they could get the government to step in and help. That's why so many farmers lost in that time, because of that.

CHESTER: This one loan officer that was a pretty good friend of Ivy's brother, he says, "We're the bad guys, the loan officers. Because," he says, "we gave you the loan—maybe you shouldn't have had the loan, and now we got to take the property back." But he said that's the way it works.

KATE: Did things have to happen the way they did?

CHESTER: Not really. I just feel that if the loan wouldn't have been so generous—if they would have kept stable and not looked to the government for a handout, maybe all the farmers would have been in better standing than they are today.

The Van Hoffs add an important twist to the private discourse of blame that punctuates public accounts of the farm crisis. Themselves the recipients of top-tier credit privileges, they question the motives of lenders who appear to ride roughshod over the social norms and obligations of a farm economy based on cultural credit. To their way of thinking, these lenders were both too generous and too quick to repossess and liquidate. What could motivate this behavior, they argue, except for the values of public-sector lending and the expectation of "government handouts"? Poking up through the public story of "that's just the way it works" is the critique of a world that no longer conforms to their vision of moral order and social stability.

Jake and Dan Holquist fully appreciate Philip Van Hoff's predicament. They have been farming together ever since their parents retired in the early 1970s. In 1987 Farm Credit Services applied a new depreciation formula to the Holquists' farm and readjusted the value of their collateral. As Jake tells it, these new rules simply didn't make sense, given the cost of everything else in the farm economy:

They just—all the banks, Farm Credit, everybody—just quit borrowing money. Got to pull these loans in, everybody's going bankrupt! They all owe too much money, way more than they're worth.

And they started this devaluation process. We got a new loan officer that year, and Farm Credit suddenly come up with a new formula for evaluation. He come out and he took our $22,000 tractor and lowered it to $16,000—so all of a sudden you lost $6,000 of equity. But his formula wasn't done yet. That $16,000 was a top value, then they would cut it in half, and loan you 50 percent of that. So [its value] is now down to $8,000. And we'd just put new tires on the tractor for $4,000. So I said, "We'd really screwed up, because we should have bought some tires from *you guys*, because obviously you know what they're worth!"

Having their tractor appraised at twice the value of a set of new tires was not the end of it. When the new formula was applied unilaterally, the amount of their outstanding loan was suddenly much greater than the value of the property securing it. They were told that if they could not come up with an immediate payment to cover this deficit, all of the property held as collateral on their loan—tractors, cattle, and a large parcel of land—would be repossessed. Rather than looking for ways to help the Holquists restructure their debt, the new loan officer advised them to get out of farming. What began as a revaluation of their collateral quickly turned into a reassessment of their ability to farm. Jake is convinced other motives were at play:

[The loan officer] said, "There's no way you guys are going to make it. You owe too much money; just quit." Two months before, the president of Farm Credit come along and told me, "You guys are really an example here. Fine young farmers, you're really doing good!" And now all of a sudden we've got this complete reversal of policy. I, till this day, swear that they were going to sell out people where they could sell 'em out and break even. In other words, our loan was good enough that if they sold everything we had, they wouldn't lose any money on this. But they had some people they'd lose tons of money on. Those they restructured. They didn't restructure the smaller operators, and we were small at the time.

Like the Van Hoffs, the Holquists believe that panicked lenders were out to salvage what they could and that their farm looked ripe for the taking. Because the amount of additional collateral they owed was relatively small in comparison with other farmers, Jake and Dan were able to strike a deal: Farm Credit would give them until that year's harvest to repay their collateral deficit. After an anxious summer of weather watching, the crop came in from the field with enough profit to spare, and the threat of foreclosure was lifted—for the time being. Rather than instilling sympathy for the plight of more indebted farmers, however, this experience amplified their suspicion that some farmers were getting "charge-offs" at the expense of others.

Mr. Bankster

After a decade in the military and moves from one West Coast base to another, Roland Alber and his wife, Patricia, returned to Star Prairie in the mid-1970s. Roland had always known that when his grandfather died, it would be up to him to keep the farm in the family. Although his parents had farmed when he was growing up, they had since sold their own farm and moved into town, where they ran a small variety store. Nearing retirement, they were in no position to take over Grandfather's home place. Typical of old-name families, Roland describes a family history that serves to fix his own destiny:

> This is our family's home. My great-great-grandfather came here with his family after the Civil War. At that time land in Star Prairie was up for homesteading. This particular eighty [acres] was homesteaded in 1870. That's when [the deed] was applied for. In 1944, during the war, my grandfather bought this farm from his dad, and then in 1976 it was our turn. During and after the Korean War, Uncle Sam still needed men. So I was one of them, went to California, where I met my wife. We were living in Arizona when the farm came up to us. We had good jobs; we had no reason to come here. Well, we did have a reason, but we wasn't forced into it.

Farmers like Roland do not approach the prospect of farming as they might any other business or employment opportunity. Not only have they been groomed since childhood for the day when the farm "comes up" to them, the choice is usually an all-or-nothing proposition: either they buy the farm or it will be sold to someone else. When family members wish to divide a farm inheritance, one heir, usually the eldest son, will be given the option to buy the land and house site, thereby keeping the farm in the family.[6] This is what happened in Roland's case. However, after paying his parents and other relatives their share of the estate's appraised value, he and Patricia were left with nothing but the deed to 124 acres, a hundred-year-old farmhouse, and an assortment of aging farm buildings and outdated equipment. What began as a desire to accept the family legacy quickly became a test of character.

ROLAND: I can tell you, when we came around the corner and down the hill here to come to this place the very first time, my eyes really got damp. We'd been gone a long time and the place did not look like I had remembered it. And when you get to digging around and searching and wiggling and moving and checking, there wasn't anything here that was adequate. Nothing. The buildings were just about dilapidated. And I thought, Oh Lord, why me?

PAT: For the first couple months he walked around here and felt like his grandfather was walking with him. Even though he was dead and gone. All this family feeling descends upon you because you're sitting in the middle of this and it's *your turn* to be carrying the torch for the family.

In the Albers' case, "carrying the torch" meant taking on a substantial amount of debt from the outset. The electrical, water, and plumbing systems that had met Grandfather Alber's needs simply could not support a growing family and a modern farm operation. "Electricity had been on this farm since the thirties," Roland observes. "But we come in here with our modern appliances and we could blow fuses faster than we could make a pot of coffee." What didn't need to be replaced had to be rebuilt; what didn't meet the standards of a Grade A dairy had to be acquired and installed. With a mortgage on their land from the Land Bank, a bank loan for machinery and equipment, and working steadily with the county extension service to increase the productive capacity of the soil, the Albers were eventually able to maintain a herd of thirty-six cows on land that had previously supported sixteen. Things were running smoothly until 1987, Roland says, when the Land Bank suddenly called in their loan:

In the early eighties, for what we were trying to do and trying to get established, I thought we were doing pretty well. Then 1987 arose, like one minute past midnight—just about down to that fine line. Because one month, as terribly inflated as it was, your land is worth $1,000—and *whoop!*—change the calendar, and it's worth $300! Eighty-six versus eighty-seven, *bingo!* Based on this inflated value, we had been able to borrow money, build new buildings, buy machinery. Do all these things. When you have collateral in real property of $1,000 an acre and—*boom!*—that thing goes to $300, you have been had, on paper. Because there's no way in the world you can generate enough collateral to take care of your existing debt, and should you need more money, well, only a fool would grant it to you. So when you take your total assets and add it up against your existing debt and it didn't balance, you are broke.

Confronted with the Land Bank's demand for additional collateral, the Albers consulted a lawyer. He advised them to make the interest payment on their loan but nothing toward the principal. In six months, when the next payment was due, they could take stock of the situation and decide how to proceed. The advice came from a reputable attorney who had worked with enough distressed farmers to know that "buying time" was often the best, if not only, strategy at their disposal.

"The man that they think is their close, personal friend in the bank—when push comes to shove and times get a little tough—he's not really a close, personal friend," says Roland Alber. "Mr. Bankster looks out for number one, and phooey on everybody else." When the Federal Land Bank and the local bank moved to foreclose, they made arrangements to hold the auction at the farm site. But Pat and Roland had learned through their legal advocate that they were under no obligation to "host" such a thing. The auction was postponed, and the Albers took heart from this encounter, realizing that they were in a high-stakes game of cards, where bluffs could be called—and a poker face has its uses. During the mandatory mediation period, Roland says, they all but *dared* their lenders to take the farm:

The real problem that caused the crisis in rural America was the fact that the banks was passing money out as fast as they could fill the drawer. They did it at home; they did it abroad. They really took a licking abroad and everybody knows that. But you mustn't talk too loud about them doing it at home, because ears listen and they're close by. That's the John Gospel truth. So we go through this mediation period, we buy this time, and when we come out of it, I could see it's pointless to ask for an extension. This isn't the only farm in the world—it's a sentimental farm, but it's not the only one. They're not the only bank. Those cows, I like those cows, but they're not the only cows in the world. And somebody around here is gonna take a real licking. Oh, maybe they'll redeem five cents on a dollar, *maybe*. So if their motive is to sic the wolves after me—*woof, woof,* and let 'em come!

Before sitting down to mediate, Pat and Roland hired an independent appraiser to take a full inventory of their property. This enabled them to put the following facts on the table: their total debt was $250,000, and according to the appraisal, the current market value of their land and equipment had fallen over 30 percent and now stood at about $80,000. By the time they claimed the exemptions permitted under bankruptcy law, their lenders would be able to recoup only a small portion of the original loan. This, combined with the cost of holding the auction elsewhere, made seizure of the Alber farm decidedly unappealing. Be this as it may, no progress toward a settlement was made in mediation, as both the local bank and the Land Bank considered the Albers' appraisal "biased" and refused to accept it. The mediation period came to a close without a resolution, and by law the lenders were now entitled to proceed with foreclosure. Roland and Pat stuck to their guns and, as threatened, filed for bankruptcy. However, this process came to an abrupt halt once the lenders sent their own representatives out to appraise the farm. Roland delights in the memory of their visit:

In those days I was cultivating corn with a tractor that wasn't worth more than about $700 at the very best, and I had a cultivator on it that wasn't worth more than $75. And darn proud of that thing, did a nice job. I recall this [representative] saying, "What are you doing?" And I said, "I'm cultivating corn," and he said, "With what?" And I said, "With a tractor and a cultivator." This is the eighties, and you better wake up if you don't know how you cultivate corn out here—I mean, we ain't doing it with horses! At any rate, [he says,] "Where's your tractor and cultivator?" I said, "Geez, it's sitting right there by your car, can't you see?" He said, "*That's* what you use?" He and the lawyer look back and forth, [wondering] how in the world did this bank ever lend this kid all this money? How are we going to squeeze even a fraction of it, when he's got two pieces of equipment out there that isn't worth close to a thousand dollars? Something's haywire in Denmark here somewheres! And it sure as the world was.

After several months had passed and no legal action had been taken, the Albers arranged a meeting at the bank where they put forward a plan to restructure their loans, adjusting the principal and interest to reflect current market values. In short order, papers were signed—and, with smiles and handshakes all around, Pat and Roland walked out of the bank the new owners of their own farm. "You could have knocked us over with a feather," Roland says, recalling their amazement. But they had won.

The Albers would make an unlikely pair of bank robbers under any circumstances. But by a curious twist of logic, they are—in the imagination of Star Prairie's lenders, at least—a living incarnation of the Bonnie-and-Clydes of the farm crisis. As the recipients of loan write-downs at Farm Credit and the local bank, they are precisely the sort of farmer who stands accused of raising the cost of farming for everyone in the community. Yet Pat and Roland's experience allows us to see another side to this story. Beneath the public rhetoric of friendship and common interests lies an ineluctable fact of market capitalism: when the economic system operates to produce a "loss," it will be taken out of someone's pocket. But whose pocket that is—the debtor's or the creditor's, the community's or the taxpayer's—is a subject of ongoing political struggle and moral debate. In this sense, the primal scene of capitalist society is a matter of competing cultural perspectives, not inherently fixed social roles.[7] The ability to identify your private economic interests with those of the larger "public good" is as possible for those who have money to lend as it is for those who can only borrowers be. During the farm crisis, lenders and farmers both claimed to be on the side of the common good. Yet much as in the Farmers Bank robbery during the depression, it wasn't always clear to the community who was robbing whom.

"Take care, your worship," said Sancho; "those things over there are not giants but windmills, and what seem to be their arms are the sails, which are whirled round in the wind and make the millstone turn."

"It is quite clear," replied Don Quixote, "that you are not experienced in this matter of adventures. They are giants, and if you are afraid, go away and say your prayers, whilst I advance and engage them in fierce and unequal battle."

—Cervantes, *Don Quixote*

5 White Crosses

The idea came to Ray McClelland while he was cultivating corn on the land he was fighting to save. Always before the tractor-cades had gone from west to east—from the farm gates to the capitol steps. Always farm groups had gone to the government to demand recognition, but wasn't there a way to go directly to the American people? A way to show solidarity with the working man and woman—a way to say that this country needs its factories *and* its farms? It was like a dream, Ray recalls, this vision of what he could do. Why not go from the Great Plains to the Rio Grande? Why not travel through the heartland and make the media pay attention to what was *really* happening to farmers?

The plan that took shape in Ray's mind was a bold but simple one. He worked out an itinerary that allowed him to link up with members of the grassroots farm movement and visit a farm that had been foreclosed on at each stop along the way. Driving approximately a hundred miles a day—about all you can do on a tractor—would put him touch with about two dozen people. If the media followed him and did a profile on each of these families, they would see for themselves what the farm crisis was all about. A local International Harvester dealer loaned Ray a tractor and the Farmers Union donated a set of tires. A good friend who was also in foreclosure made the trip with him in a tractor of his own, and farmers along the route signed on for shorter legs of the journey. In January 1985, with high hopes and a sense of mission, the small caravan from Star Prairie set off on a slow 2,000-mile trek to Texas.

For Ray the trip was deeply affirming, even if it did not attract the media attention he hoped. There was something about making a personal statement, and making it publicly, that charged him up and made the effort worthwhile. As a state coordinator of the recently organized American Agriculture Movement, he wanted to show allegiance to the farmer, not to any of the "official" farm organizations that purported to represent farmers' interests. His vision of unity transcended partisan politics, and his tractor was outfitted to demonstrate this:

We had that baby all set up with flags, American flags on the front and we had an American flag on the back and a Minnesota flag on the back. And then we had Farmers Union and NFO [National Farmers Organization] signs on there too. My point was, *You all suck,* all of you. You're *stupid.* A farmer's a farmer and I don't give a shit about all these organizations! Let's get our heads here and all come together in one agreement. Well, that's not going to happen, because this damn Farm Bureau is politically one way and Farmers Union is politically the other. But that was a big dream I had, that you bring all those together. And that was a pipe dream, I guess. But I had all the signs on there, because it says I don't care who they are. There's farmers in trouble and they're *all in trouble.*

In the late 1970s and early 1980s, distressed farmers all across America were coming to the conclusion that their problems had a common source: the federal government and the Farmers Home Administration. Confronting an adversary such as this, they realized, would require something more than politics as usual. Much of the impetus for grassroots organizing came from FmHA borrowers like the McClellands who were denied credit or forced into foreclosure by the "lender of last resort." In one state after another, farmers filed lawsuits against the secretary of agriculture, contending that Farmers Home, an agency of the U.S. Department of Agriculture, was illegally foreclosing on their farms. In 1983 these cases were combined to form a class action lawsuit that involved over a quarter of a million farmers nationwide.[1] In November 1983, Farmers Home was ordered to suspend foreclosure action until it developed regulations allowing borrowers to defer loan payments in the event of uncontrollable circumstances and for a neutral mediator to hear their case before liquidation procedures could begin. Hailed as a victory by farm activists, the ruling was strenuously resisted at all levels of the agency. A legal stalemate prevailed for five years, and a moratorium on FmHA foreclosures remained in effect until 1988.[2]

Thus, when Ray McClelland set out on his solitary tractorcade in 1985, it was with a sense of personal and political urgency. Minnesota had yet to legislate mandatory farm credit mediation, and the U.S. Congress was under pressure to deal with the growing number of rural bank failures and the nearly insolvent national Farm Credit System.[3] Regional advocacy groups began agitating for a *unilateral* moratorium on farm foreclosures, one that would apply to all agricultural lenders. In January 1985 Groundswell rallied ten thousand supporters at the Minnesota State Capitol, threatening to impose a "people's moratorium" if lawmakers failed to enact an emergency measure to stop all foreclosures within ten days. The McClelland farm was scheduled to be sold at a sheriff's auction in February—and when it became apparent that the state was not going to act, Groundswell took its fight to Star Prairie.

No Sale!

As in other foreclosure actions, Groundswell's plan was to assemble peaceably and, by discouraging bids during the auction period, block the sale of the McClelland farm. Less than twenty-four hours before the event, Travelers Insurance Company postponed the sale, urging the governor and state legislature to enact a plan that would help "all distressed farmers in Minnesota." When the McClellands' sale went forward in March—and still no legislative action had been taken—five hundred Groundswell supporters rallied at the steps of Star Prairie's county courthouse. Carrying white crosses to represent farms that had been foreclosed upon and chanting "No sale!" during the auction, activists successfully stopped the sale and forced a second postponement.

A week or so before the third auction, Ray received a phone call from the Reverend Jesse Jackson, who inquired about the upcoming sale and offered his support:

My girl at that time was only eleven and she had known about Jesse Jackson, maybe heard us talk about him or maybe read about him in school. I talked to him just a brief thirty seconds, and she was standing there pulling on my arm, and I said, "I've got a daughter here that wants to talks to you." She picks up the phone just like it's her job to do and she says, "Hello, Mr. Jackson, I want you to come and help save my daddy's farm." And so—he came. I got back on the phone and he said, "Well, I wasn't going to come for you, but I'll come for her."

As in the earlier Groundswell protest, the intent of the Jackson rally was to prevent the sale of the McClelland farm. But as before, just prior to the auction, Travelers called it off—this time citing a dearth of local law enforcement and a "concern for public safety." Although the objective of the protest had, in a sense, disappeared, the rally went forward nonetheless. To the consternation of some older residents, schools were recessed for the morning to allow children the opportunity to hear Jesse Jackson speak at the city park. Afterward, the crowd marched solemnly down Main Street to the county courthouse, where Jackson delivered a memorial service for farmers in the area who had committed suicide. Once again, white crosses were placed on the courthouse lawn.

Although Alice and Ray had braced themselves for the

possibility that their foreclosure would become a highly visible affair, having a presidential candidate spend the night at their house and appear in public on their behalf was more than they had bargained for. But they remain philosophical about their hour of fame. They understood that 1985 was a decisive year for the farm movement and that their foreclosure was part of something larger than themselves. Across the Midwest, increasing numbers of farmers were looking for a collective solution to their problems, and the efforts of regional farm groups were finally receiving national attention.[4] Had the McClellands' foreclosure come a few years earlier or later, it might have passed with no more publicity than a notice in the county newspaper. As it was, however, their ordeal became a local lightning rod for a crisis that was as much cultural as it was economic.

Frank Tostrud's office at the Star Prairie Bank faces the corner of Main and First Avenue. From this vantage point he was able to observe the people who gathered to watch the Jackson rally as well as scan the ranks of those who participated. A longtime resident and prominent figure in the community, he prides himself on being able to say hello to most of the people he meets. "If they're farmers," he says, "I probably know 90 percent of them by name." Yet, as he tells it, there were few familiar faces on the street that day:

Well, when Jesse Jackson walked through town, there weren't very many people we knew in the crowd. They were all brought in from outside. It was a media show. They were largely people that had failed before this thing came along. Not totally, but they were the ringleaders, the ones who got people worked up. So it wasn't something that the—what we call *real people*—paid much attention to. It wasn't an appropriate way of addressing the problem. March in the street and talk about—I don't know what they talked about. I didn't go to that thing.

Community bankers were not the only ones who dismissed the rally and the protesters as "unreal." As Lester Nordman, a public accountant with many farm clients, observes:

The people that were involved in the tractorcades and so on, the Jesse Jackson [rally], and the protests on the courthouse lawn, and stopping the sheriff's auction and so on, at the time were viewed as very radical. A lot of them had gotten overextended and it was kind of a case of sour grapes. That was the general consensus. When I look back at it, I guess I still feel the same way, when I look at the people that were involved. You didn't see a lot of the farmers that were having problems dealing with their problems in that way. Many of them were the ones that would go to the bank and work it out as best they could, or go to the lenders and say, "It's not working." They had to give to get. It seemed at the time—many of the ones that were doing the protest didn't want to give to get. They just wanted to get. After they started protesting and taking some of the actions that they took, I think that only reinforced their image as "not one of us," so to speak.

The assertion that protesters were "outsiders" who acted in "radical" or deviant ways is a staple part of Star Prairie's public account of the farm crisis. Few respectable members of the community wanted to be seen or "caught dead" at the Jackson rally, and those who were there often claim to have had other reasons to be on the street at that time. Some of the community's negative reaction to the event stems from a thinly veiled racism that would have found Jackson's presence in town alarming no matter what the political cause. In a community where "ethnic diversity" refers to having ancestors from Germany or Norway, prejudicial beliefs about African Americans exist largely unchallenged by everyday experience. Thus, it was not uncommon to hear Jesse Jackson dismissed as "that nigger" whom no one "wanted to see anyway." But racist attitudes were only part of the story underlying the community's aversion to the courthouse rally. Of greater importance is the sense in which protesters flew in the face of social convention. Unlike hard-pressed farmers who privately restructured their loans, protesters were perceived as wild-eyed radicals looking to get something for nothing.

To be seen as willing to "give to get," farmers had to do

whatever they could to meet the terms of their original contract. This might involve putting up previously unleveraged assets as additional collateral or selling land or machinery to pay down a portion of the loan. In exchange, the bank or deed holder might extend the contract period or lower the interest rate. If none of these concessions were possible, then farmers were expected to accept repossession or foreclosure as the lender's legal right. Ivor and Helen Lindstrom speak for many farmers in Star Prairie when they say that public protest was not the "right way" to deal with financial problems.

IVOR: Whether it's conservatism or whatever it is, when it came to the actual demonstrating at the courthouse or trying to stop legal proceedings when the farm was being auctioned, most of the people in our community felt that this was an obstruction of justice and therefore it was wrong. Even though the empathy may have been there for the person who was losing it. Yet this was not the right way to go about it. In so many cases, professional activists from the Twin Cities area would come into the community and stir it up. They were able to guide the thinking or the actions of our local people and get some of them to react in ways that they normally wouldn't have.

HELEN: We also had a couple of people from our community that were involved in that that were professional agitators themselves and would also go out to other causes and work with them. So we had local people that were really in the lead of that, besides outsiders that came in. They also were outsiders that went other places.

The specter of "outside agitation" made the illegal behavior of protesters all the more threatening, as it appeared to cause members of the community to do things "they normally wouldn't." It is as if the impulse to protest were both alien and contagious—a way of thinking and acting that could throw an otherwise orderly society into turmoil. When I conducted interviews in Star Prairie, I met fewer than half a dozen farmers who admitted to participating in the courthouse protest. This small number may be due to the out-migration of dispossessed farmers over the ten years following the peak of the crisis, or it may reflect the desire of some to downplay their involvement after the fact. Whatever the explanation, it is extremely likely that the majority of those who took part in the Jackson rally did, in fact, come from other parts of Minnesota, as well as from nearby regions of Iowa and the Dakotas. Groundswell, PrairieFire, and American Ag were then in a position to mobilize a far-flung network of farm activists in these states, and the prospect of appearing in the national spotlight undoubtedly galvanized the faithful.

As prominent representatives of the protest movement, the McClellands achieved local notoriety as "ringleaders" and "professional agitators." This was not an enviable position to be in, however, as dairy farmers Boyd and Nancy Martin make clear. Born and raised in the house where we sit talking on a late summer evening, Boyd bursts into laughter when I ask him about the courthouse protest.

BOYD: Ray McClelland was an idiot! He never did know how to farm and he never will know how to farm. He blew up that thing in such big proportions. Sure, there was a big problem with farming. Of course, he's one of these guys that likes to have a lot of—I can't think of the word I want.

NANCY: He has to have the newest, the biggest, the best equipment all the time—a new pickup truck.

BOYD: Yeah. The biggest culprit of that whole business was Ray McClelland. He's a guy that likes to have excitement and likes to stir up trouble and he's always been that way.

NANCY: A rabble-rouser.

BOYD: There are a lot of them around. He just *likes* to do stuff like that. He was not really—his problem wasn't related to what was going on in the farming industry at the time.

NANCY: He was just a very poor manager.

BOYD: A lot of good farmers knew what to do and tightened up their belt and made it. He just never could do that.

Within the community, even those who have never met Ray have a negative opinion of him and often take pleasure in passing along gossip, no matter how far-fetched. Craig Borstad, a local businessman, offers these gratuitous details:

Ray McClelland is a perfect example of somebody that had his livelihood threatened and was like a caged animal, so to speak. He probably spent more time campaigning or telling everybody his problem, than he spent farming. He really didn't want to admit defeat or admit fault on his part. I don't know that that many [people] agreed with him. He doesn't, as far as I know, have very many close friends around here. I was told that before he had his problems, he milked for a while, and a fire burned down his barn. The neighbors came and helped clean up and before it was all cleaned up, he went off on a trip to Hawaii! [*Laughs.*] Well, you know, the neighbors don't come again. I've also been told, and I can't prove this, that a lot of his problem was due to his losses in the commodity market, not his farm.

The McClellands know that their participation in the protest movement caused more than a few raised eyebrows. In the glare of public scrutiny, their financial troubles were attributed to every form of mismanagement conceivable. At the height of the crisis, the harassment they suffered forced them to take their children out of the public school district and enroll them in a private Christian school some distance away. When they try to explain the community's response, they attribute it to the fact that they are still considered "newcomers" in Star Prairie, even though they have lived here for twenty-five years.

ALICE: Who was it that said, "It's the right cause but the wrong people fighting it?"
RAY: That was the school administrator.
KATE: What did he mean?
RAY: We were the wrong people to fight the battle. It should have been somebody else other than us.
ALICE: We had moved in here, is what that meant.
RAY: The banker in town, his comment was, "You earn a penny, you save two—and McClelland didn't save any money." Other farmers made comments like, "McClelland does some erratic

> things," or, "He spends too much money and does this or that. He bought too big of a tractor." I bought a Ford and it *was* a big sucker. It was a big pretty blue one and they weren't used to seeing that around here, so that made them mad, evidently.

In response to this local condemnation, Ray and Alice point out that they had to start farming without the financial help of their families and therefore had to borrow to purchase what they needed. In this sense, the McClellands were arguably quite representative of the type of farmer likely to be caught in the crisis nationwide: young farmers with limited capital who took advantage of optimal credit conditions in the 1970s.[5] Advocacy groups unequivocally championed farmers like these, adamant in the belief that those of modest means should not be shut out of farming. Why does the community reject them as being "wrong" for the cause, when by national standards, they were so "right"?

Welfare Mentality

A. R. Radcliffe-Brown's classic analysis of joking relationships has bequeathed to anthropology one of its enduring insights: If you want to know where the fault lines of a community lie, listen to the jokes that people tell.[6] Interestingly enough, most of the jokes I heard in Star Prairie have something to do with farm subsidies.

Q: What do you call a basement full of farmers?
A: A whine cellar.

Q: Why was the farmer buried in a shallow grave?
A: So he could still get a handout.

Q: How do you make a farmer go bankrupt?
A: Weld his mailbox shut.

At first glance, it seems odd that farmers tell these jokes *about themselves*. Stranger still is the genuine mirth that follows on the heels of such a joke, no doubt told a thousand times. Why do farmers laugh? Are the jokes really "true"? Or do they refer to "farmers" who are somehow not like the teller of the joke? The answer, I believe, is a little bit of both. Yes, farmers must depend on price-support payments, but not because they believe the system is a good one. Just as young men in a matrilineal society may joke about their maternal uncle in his presence, farmers joke with one another about their thankless position in capitalist society. In both cases, the jokes are less about the individuals involved than they are about the social system that puts the teller of the joke at a structural disadvantage. Thus, like the joking relationship in a kin-based society, subsidy jokes poke fun at hapless farmers trapped in a system that is not of their own making. Would that things were otherwise, these jokes say. Their humor comes from a sense of resignation—and tacit acceptance of those who make the best of a bad situation.

In this cultural context, farm activists are clearly lacking a sense of humor. Their demand for higher subsidies makes them the butt of a joke that quickly ceases to be funny. When potato grower Cyrus Voigt expresses disgust with the protesters, it is the "whiny farmer image" he cannot tolerate:

> I never got into that stuff [protesting] because there was too much of a whiny farmer image as far as I'm concerned. Any time you stand around and complain on TV or the radio that things aren't fair—well, you know, you could say, go get a different job! Just quit and go do something else then, if you don't like it and it isn't fair. People that weren't from on a farm, when they watched farmers going to Washington and standing on the courthouse steps complaining, I think they got the wrong idea that all we did was complain. Some of the people that did the most squawking *shouldn't have been farming in the first place!* The ones that really were trying hard and really dedicated to it weren't there. They were home working. Period. I think there's a difference in the farmers. We're not all the same and that's why I had trouble with those protests.

The major difference that Cyrus has in mind is the distinction between farmers who work hard and tighten their belt when times get tough and those who look to the government to make their life easier. Beneath the surface of his claim that some protesters "shouldn't have been farming in the first place" lies the belief that public-sector borrowers use federally-subsidized credit in extravagant and irresponsible ways. Hog farmers John and Lisa Krueger make a similar point when they recoil at the image of farmers as "welfare cheats."

JOHN: It's how you feel about the guy that goes up to the welfare office to collect food stamps in his new Cadillac. This is an image I thought many of them [in the tractorcades] were portraying. From the standpoint of what they were demonstrating for, I was sympathetic to it. But for a lot of the individuals, I don't have a whole lot of sympathy. Because I know of one here, gosh, his lifestyle! He wanted to maintain a lifestyle of the new car, the new pickup every year, all the material things that is more or less the American dream. But when you got champagne taste and a beer budget, it gets a little tough. They were spending beyond their means, probably both for their family and their business operations.

LISA: The one comment we always made whenever we saw the media coverage of a tractorcade was, "Where could all these people get all these expensive tractors to be in this tractorcade?" We were sitting here with something that was at least ten to fifteen years older. That was the point. Were they really portraying, quote, "the American farmer"? Because, in the materialistic sense, there was always these big four-wheel drives and these huge cab-type tractors that were going through these motorcades, and it was never these little putt-putts that a lot of farmers farmed with. Those who were in the tractorcade, so many of them appeared like they had lots of time to be there, whereas, quote, "the average farmer" [thought], Where would you find the time to be able to participate?

Simply put, onlookers in Star Prairie were never convinced that protesters represented farmers like themselves. Their politics, lifestyle, and behavior in public marked them as people who thought they could achieve the consumption standards of the American dream without having to work for it. To local sensibilities, farm activists represented precisely the kind of farmer who made farming difficult for everyone else. Harlan Wade, a prize-winning cattle breeder, puts it this way:

There was a lot of noise and Jesse Jackson was here protesting. I didn't even go over there, to be honest. We grew up with a work ethic and with parents who had to work hard for what they had. A lot of people that protest a lot of this really aren't very good managers. Some of the biggest advocates and noisemakers were the people who never paid their bills before they got in trouble, or didn't get along with their neighbors, or were poor farmers. That kind of turned me off. It was almost embarrassing. A lot of it was, these tractorcades—they were driving a $200,000 tractor to protest that [*laughing*] they *weren't making enough money!* Well, the guys that were out working hard and driving a [less expensive tractor] and getting the job done were surviving. But the people that had to have brand-new everything and the biggest and the best—and expanded to the point where they were driving smaller farmers out of business—they were the ones that were screaming the loudest. The people sitting on the side of the road said, "What are they bitching about? Their tractor's worth more than my house is!"

Perhaps it is ironic that most television viewers are not able to tell the difference between a new tractor and an old one, and that there was probably more sympathy for protesting farmers in the nonfarm world than there was in their own communities. But the cultural tension underlying the distinction between "big" farmers and smaller ones is very real in places like Star Prairie, where, as Harlan observes, small farmers are in constant danger of being forced "out of business" by larger ones. The twin perceptions that protesters were "larger" than the average farmer and that "big" farmers benefit the most from government subsidies are not without basis in fact. Sociological surveys have suggested that farmers who engaged in protest had significantly larger farms and sales volumes than those who did not.[7] Moreover, although farms of all sizes are eligible to participate in the government's commodity programs, price-support subsidies do not benefit all farmers equally. A disproportionate share of the benefits goes to producers who are considerably larger than average, such that it would be a mistake to think that these programs were specifically designed to support the family farm.[8]

The farm crisis threw the cultural boundary between "big" and "small" farmers into bold relief. Populist rhetoric was employed on the side of protesting and nonprotesting farmers alike, yet it was only among those who sat on the sidelines—watching the parade of media images—that the populist critique of big government and big business was extended to include farmers themselves. Sociologist Kai Erikson has given us a way to think about conflicts that divide a community from within, setting neighbors against one another at what can seem the merest of offenses. At stake in such crises, he argues, are the contours of the community itself, as members struggle to define the moral boundary that separates insiders from outsiders and friend from foe. To mark such a boundary, a community seizes upon what is ready to hand: the behavior of its members. As Erikson writes, deviant behavior constitutes "one of our main sources of information about the normative outlines of society. In a figurative sense, at least, morality and immorality meet at the public scaffold, and it is during this meeting that the line between them is drawn."[9]

On the public scaffold during the farm crisis was a confrontation between farmers who managed to make do with what they had and those who insisted that what they had was not enough. When residents of Star Prairie looked closely at the faces of those involved in the protest movement, they marked as deviant one they knew well. It was the face at the bottom of the credit structure: that of a Farmers Home borrower, the farmer who had been given "something for nothing" and foolishly squandered it—yet had the audacity to demand more.

Wendell Norris, now a local bank consultant, recalls his earlier days as field agent with FmHA and puts two and two together:

When I worked for FHA, I could kind of see some of these guys going that route [of becoming politically active] just because of the financial shape they were in. I remember seeing several of those people I visited on my FHA visits in eighty-five to eighty-six being involved with the courthouse scene, and really being the leaders of that farm group that was protesting and going for mediation and trying to get their homesite carved out of their real estate mortgages and things like that. [The county supervisor] and I talked about that, and he knew what was happening with FHA. He knew the people that were in the foreclosure mode at that time, and it was a lot of those people that were involved in that farm movement.

Joe Clark, a former farmer and current director of Farm Credit Services in the Star Prairie region, is convinced that much of the farm problem in the 1980s came from Farmers Home borrowers who simply didn't know how to manage the money they were given. In his opinion, FmHA's low-interest, low-collateral loans were essentially an entitlement program:

I look at FHA as a form of welfare. They just put the money out there—and they *kept putting* the money out there. These young kids had—I mean, I farmed with kids out there that had *ten times* the equipment I had! You know, I'm sitting there saying, "I can't even afford to buy a car, and they're driving brand-new four-wheel drives and whatever." There are none of 'em farming today. Part of it is because they had no responsibility for what they did. Why should I care, [they thought,] if I don't have to pay it back? If I lose it all? I got nothing in it. I didn't put any investment in it to start with. So if you've got nothing to start with, then why are you going to have any pride? The majority of them had no pride in what they were doing. And it just went amok, to me. That's where the farm problem came in. They just had no responsibility for what they were doing.

A fundamental suspicion of federal programs links the privately expressed concerns of local lenders and farmers. Bonnie and Kurt Holloway are an instructive example of farmers who eschew government assistance all together. Their small dairy farm is able to support itself, but their household expenses are paid for with the income Bonnie earns as a registered nurse.

Kurt and Bonnie acknowledge that they could expand their farming operation and have more to live on if they applied for an FmHA loan and enrolled in the federal commodity programs. But this is something that Kurt, a Vietnam veteran, is unwilling to do.

KURT: When I come home from the [war], I was never going to get involved in another "government program," and I never did. That's just it.

BONNIE: I think Kurt carries enough scars from Vietnam.

KURT: I just walk in the dang building and they're all built the same. They don't have any pictures on the walls, got a bunch of desks in there, with people shuffling paper. I just get spooked. When you go out of [government buildings like that], you get shot at. Just like [it was] in Vietnam, it's just a game to make money for somebody. I think that's what the government programs do. Even in agriculture. They are nonproductive. They mess up the market structure.

BONNIE: One thing that Kurt and I have had some gut feelings about is we've kind of watched our neighbors and it gets real frustrating. Heck, here we're—coming along, but, I mean, they'll drive new pickups and new cars, and you really have to question, Do they just owe that much debt? Do they just go out and buy it and owe that much on it? How are they financing these payments? It just does not make sense to us. The best we can figure out, the only real inkling we have, is that maybe the government payments is what the difference is. So they're literally living off the government payments.

The Holloways are not hard-core militia types ready to jump to hasty conspiratorial conclusions, but they do give us a feeling for the cultural substrate upon which such movements are built. Not only is there a sense that government programs are "nonproductive," they appear designed to benefit people who are nonproductive. Thus, even farmers who enroll in commodity programs feel compelled to express disdain for those who cannot farm without *more* government "handouts." Like cattle farmer Oscar Anderson, most of Star Prairie's established farmers advocate doing away with "subsidies" all together, and they see this as the key difference between themselves and farm activists.

OSCAR: It seems to take what you might call a radical movement such as that [in the 1980s] to bring the attention to something. This is what brought about the mediation [program]. So something good did come out of it, something very positive. But, yet, the attitude of a lot of those people was, you know—they were back out there again with their handout. Give me something without working for it. These subsidies did a lot of that.

KATE: Aren't most farmers dependent on subsidies, whether they want to be or not?

OSCAR: Yes, definitely. I am myself. We are. They've put it into the cash flow and how do you get it out? There's going to be a hurt there to get rid of it. But if we could get rid of it once and for all, then we would all be better off by far.

It is worth noting that Oscar was the recipient of a large loan charge-off at the local bank as well as from a national insurance company. That he feels able to point the finger at those who sought government support—without implicating himself in their "welfare mentality"—speaks volumes about the sense of entitlement that private-sector borrowers maintained during the farm crisis. By claiming they can farm without "the government," recipients of cultural credit act to reinforce the cultural boundary that separates them from those who depend on federal assistance: rich farmers and poor ones. This "middle-class sensibility" comes with an emotional cost, however. Based on the public presentation of a "class" identity relative to those above and below, it requires the constant surveillance of others' economic behavior and relentless vigilance over one's own.

Dairy farmers Chuck and Sara Nugaard give voice to this uncertain class location when they express discomfort with the Farm Aid music concerts, organized at the peak of the farm crisis to raise funds for distressed farmers.[10] As the Nugaards saw it, this public cup-rattling was an insult to their personal and professional integrity.

CHUCK: I always wondered about all these Farm Aid concerts, whether we were looking like we needed handouts all the time. Or do we need some concrete changes to be made, rather than—don't give us a Band-Aid! Teach us how to not get cut next time, that sort of thing. When you give somebody something for nothing, that's what it's worth: nothing.

SARA: Yeah. And I felt like maybe the country was looking at us like, "Poor dumb slobs, they can't even help themselves, we've got to help them out." I don't want people to feel that way. I want them to respect what we do. That's my sermon for the day!

At the national level, the farm movement of the 1980s drew attention to what was happening economically in the heartland. But local anxieties were focused on another kind of crisis. Stepping into the cultural space that separates "average" farmers from "rich" beneficiaries of commodity programs and "poor" recipients of federally subsidized credit, activists called for aid to the "family farm." But whose side were they on? Their apparent allegiance to the rich *and* the poor—how can someone with a $200,000 tractor claim to have no money?—made them representatives of a cause that few in Star Prairie were sympathetic to. Yet the movement was not without effect, socially or culturally. Farm activists can claim an instrumental role in the passage of mandatory farm credit mediation laws, first in Minnesota and Iowa and in the states that later followed suit. And, at a time when it seemed that economic distress had a permanent urban zip code, protesters brought rural America back into the nation's conscience.

Less tangible, but nonetheless real, are images that linger. Like the fields of white crosses scattered along the coast of Normandy, the spare wooden crosses that stood for a time on the courthouse lawn speak to the ages. There is something here, they say. Something to mark the presence of those who are now absent; something to commemorate the spirit of what is no longer possessed, but eternally cherished. The white crosses of the farm crisis serve to remind us, I think, of the nobility of taking up a battle known to be lost before it is ever begun.

Figure 6 A moment of silence for farmers who committed suicide. Courtesy of Tom Dodge.

ERSHIP
OW
WANT A TIONS
NOT AUC IONS
MOR
MORAT
MORATOR

Figure 7 Reverend Jesse Jackson leads march down Main Street. Courtesy of Tom Dodge.

Figure 8 Protesters in front of the county courthouse. Courtesy of Tom Dodge.

Figure 9 White crosses on the courthouse lawn. Courtesy of Tom Dodge.

There was, it seemed, no person in town who was not living a life of shame except Mrs. Bogart, and naturally she resented it. She knew. She had always happened to be there. Once, she whispered, she was going by when an indiscreet window-shade had been left up a couple of inches. Once she had noticed a man and a woman holding hands, and right at a Methodist sociable! . . .

[Carol] stopped on her own porch and thought viciously: "If that woman is on the side of the angels, then I have no choice; I must be on the side of the devil. But—isn't she like me? She too wants to 'reform the town'! She too criticizes everybody! She too thinks men are vulgar and limited! *Am I like her?* This is ghastly!"

—Sinclair Lewis, *Main Street*

6 Open Secrets

The road that passes the Hanson farm is a main artery to northern Minnesota, a popular vacation spot. For most of his life, Luke Hanson has spent long days in early July mowing road ditches and putting up hay. Each year, around Independence Day, the road is virtually uncrossable due to the steady stream of cars, campers, and boat trailers heading up north. And each year, as sweat rolls down his back and the holiday traffic passes him by, he concentrates on the secure future he is ensuring for himself and his family. Like the little pig who builds his house of bricks, he feels he will have nothing to worry about when the wolf is at the door—and no reason to feel sorry for those who opt for leisure instead of labor.

When Luke and his wife, Jody, think about farmers who had it rough during the farm crisis, this is the image that comes to mind. Like most of Star Prairie's established residents, they are inclined to believe that farmers who lost their farms were not very good farmers to begin with. In their view, the surest indication of bad farm management is the attempt to live higher off the hog than everyone else.

LUKE: The few that were involved in [the foreclosures and the protests] were sort of the borderline farmers in the first place. They really weren't that good a farmer. They were ones that thought they were going to be grandiose and go for the big farms with the nice shiny new equipment. It all looked good on paper until the escalating of interest rates and the downfall of the prices. Then it sure didn't look very good. When they went into it big, it was a hard pill to swallow. I guess I really didn't feel that sorry for 'em. It's a hard line to take, but I learned early on, from being out bailing hay and watching people drive by on the weekends with their boats and campers, and thinking, Boy, that'd be fun to do.

JODY: It was hard to band together with them to bail them out for maybe living too high of a lifestyle earlier on. I guess if they wouldn't have been as flamboyant with the equipment and the lifestyle that maybe we'd have thought, They don't deserve to go under. But, that is a really hard line—to say, "*Maybe they deserved it.*" But a lot of the first farmers to go bankrupt were the ones that were strutting around with their big four-wheel-drive equipment saying [to other farmers], "How are you ever going to get done with that four-row planter? How ridiculous! You must not be doing very well, since you don't have this nice machinery. You're just plodding along."

What is it about "strutting around" and "big equipment" that makes a farmer deserving of failure? Or, conversely, why should "plodding along" and "small equipment" make a farmer deserving of success? Given the historical trend toward larger, more capital-intensive farms, these distinctions fly in the face of expert opinion, which equates business expansion with the "economies of scale" that allow farmers to be more competitive. Why are those who might arguably be the "most likely to succeed" the very ones to invite contempt?

Within local farming culture, as Luke and Jody Hanson observe, a "hard line" separates good operators from bad ones.

To be worthy of cultural credit, as we have seen, you, or senior members of your family, must demonstrate the qualities of character that uphold the social order. But Star Prairie is no place to sit back and rest on past laurels. Your fields and farm are the front stage upon which the assessment of your "backstage" morality is continually based.[1] Crooked rows and unkempt yards are indisputable signs of bad management, but to be absent from the stage yourself—or to pursue a "flamboyant lifestyle" with "shiny new equipment"—is the most damaging sign of all. For it is the moral fortitude to make hay while the sun shines—that distinguishes good farmers from those who are "borderline." And the community, as the Hansons remind us, is always watching.

The Joneses

Virgil Thompson and his brother, Arthur, farm land that was homesteaded by their father's great uncles in the 1870s. Theirs is a medium-sized operation of 360 acres and about fifty dairy cattle. Both men are bachelors, and their lifestyle is spartan, to say the least. Had Virgil not waved to me from the barn when I drove up to the farmhouse, I might have concluded that it was no longer lived in. With its plain white exterior, no foundation plantings or lawn, and a sparsely furnished interior, it was, as Virgil describes it, "nothing fancy: a working farm." Not a place, I am certain, where "overspending" was likely to be a problem.

When Virgil talks about the 1970s, he expresses incredulity at the lack of restraint that characterized the lending relationship:

In the seventies you could do no wrong. Just plant the seed and get out of the way, because the crop will come up! And then you go and sell it, and the banker'll give you money for whatever you want. It just *scared me*. I'd go to the banker and say, "Geez, my old disk is getting bad, could I go to an auction sale and buy a used one?" He says, "Heck no, go to the dealer and buy a new one!" It goes to your head. There was a time when you couldn't buy enough. Overconsumption. That was back right before the interest rates went flying up to the top. It was a scary period to be paying 18 percent interest. Agriculture, when it works right, will only return 3 to 4 percent on your investment. So you have to have things going pretty good, just to hold your own. But reverse psychology works on the masses.

Agriculture's traditionally narrow margin of profit, Virgil suggests, dictates more than a prudent investment strategy: it also precludes an extravagant lifestyle. Good farmers, he says, should know this. But the 1970s induced more than a few farmers to cast caution to the wind. Why did this happen, and why were the Thompson brothers exempt?[2] The key, Virgil tells me, lies in the social pressure to consume—and in having the independence of mind to resist it. The intoxicating idea that you "cannot buy enough" *can* "go to your head." Unless you are able think for yourself, he argues, you will be led to believe that since everyone else is doing it, it must be all right. This is the "reverse psychology" that "works on the masses." Like a herd of animals rushing over a cliff, farmers were led to act in self-destructive ways, simply by following the behavior of others. "Keep up with the Joneses," Virgil quips, "and the Joneses don't know where they're going!"

In a farming community, trying to keep up with the Joneses is a risky pursuit for a number of reasons. Whether the material goods in question are acquired for productive use on the farm or for domestic use within the household, most purchases must be made on the basis of the farm's profits, over and above basic operating and living expenses. For most family-owned farms, this margin of profit is extremely narrow, even in the best of years. Any decline in farm income affects the total cash flow, and if the farm is to continue operating without debt, a portion of the household budget will have to be used for operating expenses. Unless family members work at off-farm jobs to supplement the farm's income, there is generally little money to spare for anything but the necessities. Thus, when it comes to household spending, the high-flying lifestyle of "the Joneses" is technically out of reach for most farm families.

But there is another way in which farmers may attempt to keep up with their neighbors—and here the stakes are much higher. For it is not just houses, cars, and immaculate lawns that are signs of status in this American community: it is barns and tractors, silos and livestock, irrigation systems and rolling acres of grain. To compete on this scale, farmers must continually monitor how their neighbors are making money—and what they're spending it on. In a very real sense, a farmer's success depends on keeping up with technological innovations and standards of production within the community as a whole. As agricultural economists have noted, early users of new technology gain a significant advantage over those who are slow to innovate.[3] When farmers invest in new equipment that reduces per unit operating costs by increasing production, they become more competitive and other farmers are encouraged to adopt the technology. Yet as total farm production goes up, commodity prices fall and the last farmers to adopt the technology do so long after its benefits have diminished.

More than most Americans, therefore, farmers find themselves in direct competition with their neighbors. "What's bad for some farmers is always good for others," says retired farmer Arlo Enger. "And *this* is the whole concept behind farming. If you want to survive, there's two ways of praying: you can say, 'God, let it rain,' or you can say, 'God, *don't* let it rain over there [on my neighbor's land].'" The logic of market competition is stamped into farmers' daily business concerns, and it is nothing to flinch from or apologize for. But its cultural paradox remains. Percolating up through the surface of everyday life, it deepens the mystery of what makes people act as they do—and what kind of community is possible in a capitalist society. Even Arlo has his doubts:

Farmers are competitive. We've got a better barn; we got better cows; we got better horses; we got better everything. They're competitive and sometimes they don't even like each other very well. They're jealous. They may drain their water off on their neighbor's land and inundate his crops and so on. The only time we're real is when our neighbor is hurt, then all of a sudden everything is forgotten. I can't understand how we can do this. There's something wrong in our theology. But I don't think it's any different than neighbors anywhere else. Competition, greed, a whole series of things seem to ball up together and provide us with a nasty attitude toward the guy next to us, who had, by sheer luck, gained something that we didn't gain and so we're jealous of him. Rains on his land and it doesn't rain on yours kind of thing. I don't know what it's all about. It's a lot deeper than I could ever think. But it's there.

The pressures of economic competition, which Arlo captures so vividly, are further compounded by the use of credit. For debt financing makes it possible to "mask" competitive practices and managerial problems that would otherwise be exposed to public scrutiny. Instead of tightening their belts in a lean year, farmers may borrow to cover an income deficit, creating the impression that nothing is wrong or that they did well in a year when others did not. Instead of making do with aging equipment, farmers may use credit to purchase larger, more powerful machinery, creating the impression that they are ahead of the game and know something that other farmers do not. The use of credit also makes the moral component of these decisions difficult to judge. Does Farmer Jones have time on his hands because he is lazy, or because his new tractor allows him to do twice as much in half the time? Is Farmer Smith jeopardizing his family's future by mortgaging the farm to upgrade his dairy, or will he now be able to produce more milk at less cost? Against such ambiguous signs of progress and prosperity, Star Prairie's farmers must evaluate their own management strategies and try to make prudent but progressive decisions. Looking to one another for a sense of what the "going rate" of farm expansion might be, they are keen observers of material culture and the moral choices it encodes.

In Star Prairie, farmers' use of credit is judged according to two criteria: whether it is used to exceed normative standards of consumption, and whether it is used to foster intergenerational continuity on the farm. Reflecting on the farm crisis, Arlo Enger articulates these distinctions well:

The first wave of foreclosures came from hot dogs. These are the guys that had the four-wheel-drive tractors ahead of time and just brought a new barn for much more than their whole herd was worth, and so on. They had a brand-new four-wheel-drive pickup, you know, and they went to Texas in the wintertime. They lived on credit cards, and I didn't have too much compassion for them when they went broke. That's, in my estimation, mismanagement. The next group of people are the people I felt very sorry for, because I knew a lot of them. These were people who had worked hard, had their farms paid for, and had sons coming along, and Dad said, "I'll help you. I'll mortgage my farm and I will get you both started." And all three lost everything. Those were the people I had compassion for.

The "hard line" that Arlo draws between "hot dogs" and those he has compassion for is widely shared in Star Prairie. These categories have resonance because they impose moral order on what is otherwise a chaotic and traumatizing process of economic dispossession. When confronted with the question of why some farmers failed and others did not, members of this community focus on the moral character of the individuals involved to explain what happened. In this sense, they, like many Americans, are steeped in what anthropologist Katherine Newman calls the ethos of "meritocratic individualism": rather than finding fault in the "failure" of the economic system, they are prone to locate failure in the habits of individuals.[4] Significantly, the behaviors that farmers highlight are precisely those that challenge the community's social order. The problem with a new tractor is not its sticker price per se, but that it is acquired "ahead of time"—before others in the area have ones like it. The problem with a new barn is not its cost, but that it is disproportionately larger than what was there before. Great leaps forward, in this community, threaten to ratchet up consumption standards for everyone. If your neighbor buys a tractor twice the size of yours and flaunts his good fortune by spending the winter in Texas, your own consumption choices can begin to look benighted, if not doomed, in comparison. But if your neighbor goes into debt to help his children acquire modest farms of their own, you can rest easy and genuinely wish them luck, as this consumption is in the name of family continuity.

"The Joneses," we might say, epitomize every character flaw associated with the abuse of credit. The Joneses, as Arlo would put it, are "hot dogs." Yet, a curious fact became apparent to me after a short time in Star Prairie. To hear folks talk, it would seem that the only people trying to keep up with the Joneses are the Joneses themselves. No one admits to competing with them—and, perhaps not surprisingly, no one admits to *being* them. This presents the anthropologist with something of a dilemma. Do the Joneses really exist? Or are they merely a figment of the community's imagination? How do you know a "hot dog" when you see one?

Code of Frugality

Elford and Beatrice Hagendorf are second-generation farmers in Star Prairie. After high school, not sure he wanted to farm, El attended the University of Minnesota—for four days. As he tells it, "They always start late in September and it was beautiful Indian summer weather, and I was sitting in there looking out, wishing I was somewhere else." So he came home. And looking back on that decision thirty years later, he doesn't regret it. His folks welcomed his return and made arrangements to bring him into the family operation as a full partner. Meanwhile, his eye was on the girl who called out the numbers on bingo night at the local Jaycee's. Bea was a schoolteacher in town, and for the early years of their marriage, it was her paycheck that put food on the table. Although she stopped teaching after their first child was born, it was her willingness to care for her aging mother-in-law and their elderly neighbors that paved the way for the transfer of the home place and a contract-for-deed purchase of the neighbor's farm.[5]

As beneficiaries of top-tier credit arrangements, the Hagendorfs used bank loans to make additions to their farm in the 1970s. Always buying used machinery, and carefully evaluating the risks and rewards of every purchase, they believe they survived the farm crisis because they are "conservative people."

EL: It was a tough time to be conservative because the bankers that we were working with were superconservative and yet they were at the forefront of saying, "We'll give ya more money if you want to go and do it." Really no questions asked. Unfortunately, that got a lot of people in trouble. We were very conservative, but not to the point where we didn't do anything. We were growing but at a more controlled pace.

BEA: A far slower pace than other people did. They grew rapidly and then fell off the end of the earth in a lot of cases. A lot of people that we knew [had to file] bankruptcy or sell the farm or go to work in town. We never came to the point where we were going to have to get extra jobs in town or whatever, but a lot of our friends, a lot of our neighbors went through that.

In the retelling, the 1970s appear to be a time of contradictions: bankers could be "superconservative" yet willing to lend with "no questions asked," and farmers could be "very conservative" yet willing to expand at a "controlled pace." In Bea and El's account, however, there is a recognition that tight lenders and timid farmers will not get very far in today's capital-intensive age of agriculture. Those who sit still are sure to be left behind, while those who take giant leaps forward risk "falling off the earth." The future, these stalwart members of the community agree, belongs to the slow and steady tortoise, not the impetuous hare.

When asked what made it possible for them to be conservative during the decidedly unconservative 1970s, the Hagendorfs reply that it has a lot to do with the way they grew up—learning to live "with nothing." Frugal living got their families through the depression, and its virtues have stayed with them as they grew older. Even to this day, Bea admits, "it's very difficult for us to spend money." Nevertheless, by the late 1970s, after putting up several new farm buildings and replacing old machinery, they had amassed a sizable amount of debt. "When we renewed our loan," El explains, "the figures looked too big to us to be comfortable with. So we said, no, this is it—we're not going to go beyond this point." Knowing when to say no, they insist, is what kept most of Star Prairie's farmers out of trouble.

EL: I don't think we're the only ones. There were other people around the area that have survived nicely. So we certainly weren't alone in this community. There were quite a few that didn't get caught up in the buying spree.

BEA: We still have debt from way back when, but we always like to see it paid off. We're not the kind who can let that debt keep growing. We don't need things. We don't keep up with the Joneses, and we don't *care* if we keep up with the Joneses. [*Laughs.*] It would be pretty tough to keep up with some of the Joneses we know!

KATE: So you wouldn't be the type to see your neighbor out on a shiny new tractor and say, "Oh boy, I want one of those"?

EL: [*Laughs.*] I've said it, but I've never done it!

Stressing the value of "buying used" or "secondhand" gives established farmers a way of making—and symbolically marking—a cultural distinction between themselves and those who "went too far" in the 1970s. The "value" of secondhand consumption has less to do with the economic bargain that may be involved than with the moral "thriftiness" it demonstrates. In this sense, the items used to signify thrift need not be acquired on the secondary market at all. Being content to live in a hundred-year-old farmhouse is, for El and Bea, an expression of their conservative nature. Even that pickup truck in the driveway, they hasten to assure me, was a thrifty purchase. Although it was bought "new"—direct from the factory—a year ago, it replaced the old truck they bought "new" seventeen years ago. It is revealing that the new truck requires an explanation in a way that the old farmhouse does not. For although no one is likely to misread the virtues of a "century home," the new pickup truck is ambiguous. It is in danger of being read the "wrong" way—that is, as evidence that the Hagendorfs are people who care about keeping up with the Joneses.

New pickup trucks are by far the most hotly contested items of consumer culture in Star Prairie. Occupying a "liminal" position between the family's mode of transportation and the farm's means of production, the pickup truck signals competitive spending on both sides of a farm family's balance sheet. Is it a household expense or a business investment? Is it just a fancy way to go to church, or is it a useful working vehicle on the farm? Much of its popularity no doubt stems from the fact that the outside observer has no way of knowing for sure. The ambiguity of the pickup truck is an example of a problem endemic to "reading" consumer culture in a farming community.[6] Just as the Hagendorfs' new truck required a private explanation to be read "correctly," most of the machinery and equipment farmers buy is open to conflicting interpretations.

Without intimate knowledge of a farmer's financial situation, there is really no way to know what any given purchase truly "signifies." Is it a wise investment or a showy bit of one-upmanship? Worse yet: could it be both? The jury must remain out until the consequences of that purchase have come home to roost. For it is only in retrospect, after a farmer has publicly "failed," that managerial problems and their associated character flaws become possible to identify. In the meantime, most neighbors are given the benefit of the doubt—and watched very closely. As Bea and El observe, customary respect for individual privacy gives everyone the chance, for a while at least, to "save face" when mistakes are made.

EL: I just looked around the community and I knew that some of the people were in trouble. Most of them didn't talk about it. It wasn't commonly known who was in trouble and who wasn't. You could assume, but you didn't know for sure in a lot of cases.

I always treated them the way I always had. Wave at 'em when they went by or talk to 'em when I met 'em on the street, but we didn't talk about it. It wasn't any of my business. If they brought it up, fine. I never brought it up.

KATE: Did you have any friends who confided in you?

EL: No.

BEA: It was really pretty hush-hush.

KATE: It sounds like it could be a very isolating experience.

EL: Yeah. Farmers in particular don't want to admit that they made a bad boo-boo. There's a lot of pride in this business. You work hard; you try to take care of the land; you try to have nice livestock; you try to run a good operation. To be even thought of as a failure in any way, shape, or form is a terrible stigma. Most people would rather eat it. They'd rather just tough it out than admit that.

As commentary like this reveals, social forces within the community actively prevent the public recognition of financial hardship. Embarrassing episodes will occur only if someone "brings it up," and no respectable farmer will be the first to do so. To admit that one has mismanaged—indeed, to admit that one has acted in ways that suggest an affinity with the Joneses—is more than most can do and, just as importantly, more than most want to hear. When well-intentioned neighbors like El cross paths with those in financial distress, both will attempt to pretend that nothing is wrong. But the performance must always ring hollow. Not only must the one avoid asking about "it," the other must not bring "it" up. Between the two looms a "secret" that both "know" but cannot share. As sociologist Erving Goffman has observed, encounters like these instantiate a fundamental social inequality between those who are empowered to consider themselves "normal" and those who must function with a "stigma."[7]

Neighborly politeness during the farm crisis did more than reinforce a sense of stigma, however. For the code of frugality is a double-edged sword: not only does it allow the community to make a distinction between farmers who "deserved to fail" and those who did not; it forces individual farmers to judge *themselves* according to these collective norms. Had they been beguiled by a neighbor's new tractor and acted impulsively to acquire one like it? Had they let the loan officer talk them into taking on more debt than they could realistically handle? Had they given in to the desire to provide nice things for their family? A communal ethic of frugality offers little solace to those who ask such questions of themselves. For in the final analysis, it says, a person's moral worth depends on self-control: on the containment of consumer desires and on being able to "make do" with less. In this cultural context, farmers in financial distress could be as pitiless toward themselves as their neighbors were. Social encounters therefore took on the form of what literary critic Eve Sedgwick has called the "open secret": a situation in which one person experiences the humiliation of trying to hide a discrediting fact that the other appears to "already know."[8]

The Last Resort

Jeannie Travis never wanted to farm, not really. But the military man she met at a church social in her Texas hometown swept her off her feet with promises of wealth and the beauty of rural life. When Martin Travis was discharged from the army and the young couple came to Star Prairie in 1954, Jeannie was mortified to find herself settling down among the poorest people she had ever seen. All the women appeared to be "downtrodden," she recalls, and all the children seemed to be without shoes. She knew that times were hard for farmers in the 1950s—farm incomes had slumped after World War II, just as they had after World War I—so she kept her misgivings to herself and tried to make the best of her situation.

Martin did not take over the home place right away since his parents were still living there—and counting on their large brood of children to keep the farm running. To Jeannie's consternation, he spent more time working on his parents' place than he did tending to business at home. Although other farmwives had warned her "never to learn to milk" if she didn't want to spend her life in the barn, she rolled up her sleeves and got to work, hoping that her support would help Martin accomplish his dreams. Things began looking up in the mid-sixties. With financing from the Farmers Home Administration, the Travises bought a 260-acre farm on good soil and, under the tutelage of FmHA's county supervisor, were soon operating a 30-cow dairy. Yet the prospect of better times did little to allay Jeannie's doubts about Martin's ability to manage money wisely. As farm families began to feel the effects of rising incomes and property values, the lure of consumer goods turned more than a few heads. And lenders, she says, were ready to capitalize on this pent-up demand:

"Hey, Joe," [the Production Credit officer would say,] "you're doing all right now, but if you were farming more land . . . "—that was part of the push: buy more land, buy more machinery. "You know, that old machinery of yours, why don't you replace it with new?" And of course that was just music to these farmers' ears, because here they would go out with this bright shiny tractor to make Joe Blow down the road just as jealous as can be. We could see their farms. So you could see that bright tractor out in the field, and that would get everybody so upset! By noon, everybody within a ten-mile radius knew that farmer had a new tractor. Well, what do we do but go to the bank? How did Joe get a new tractor? "Well, do *you* want a new tractor?" [the banker would ask]. Sure! And so it was playing an ego thing with men and they kept pushing it, and pushing it harder and harder.

But men were not the only ones who jumped at the bait. Farm women also had reason to demand their share of "easy money":

We ladies said, "Gee whiz here, honey, you're putting this [money] outside, but we need some money here on the inside. I need—this rug is so worn out, there's holes in it. The curtains are so worn out, I'm having to tape them together so I can pull 'em shut to keep the sun out." And the husbands would say, "You don't make money on the house—so the money's got to be spent on the outside." So that's why we didn't have anything in the house, to speak of. Finally the little lady got so tired of that, she was going to have something in the house! In the seventies, there got to be so many little women that said, "Hey now, we're getting a little guts now." And the bankers or loan companies had gotten more generous. [They said,] "Sure, why don't you redo the inside of your house? Why don't you build on some rooms?" Things like that. Which made it much easier for them to chop our heads off.

Jeannie remains convinced that there was a "deliberate plan" on the part of agricultural lenders to make farmers "go belly up." Why else would they have pressured farmers into signing loan agreements that, year after year, only pushed them deeper into debt? By 1981 it was apparent to her that the farm was not generating enough income to cover routine operating expenses. Yet in the early 1980s, Farmers Home and Farm Credit were still looking at the fact that the Travis farm was valued at $300,000, and on this basis, they were perfectly happy to extend additional credit. To make matters worse, Martin had begun to drink heavily, and as his behavior became progressively erratic, the majority of the farm's daily operation fell to Jeannie and their five children. Feeling unable to continue farming like this, she appealed to the FmHA supervisor to help them get out of farming while they had assets left. But at this point, neither lender had any interest in liquidating the farm, and Martin was dead set against it.

In 1981 and 1982, Jeannie felt that Production Credit "coerced" her into signing new loans for the farm's annual operating expenses. Martin had taken to leaving the farm for days on end with no explanation and had become verbally and physically abusive when at home. Jeannie despaired of ever finding a way out of an increasingly desperate situation:

Kate, I was the best actress. Everybody thought I was so in love with the husband. But ten years ago I fell out of love with that husband. I was living with him because I was going to get those kids educated, over my dead body. One winter day, we were in the barn.

He come up the runway with a pitchfork, and I thought he was going to *shish-kebab* me with a pitchfork. I told him, "You pitch that pitchfork!" He put it down and then he come with his fists and started beating me. He beat me until I passed out. I woke up—he was in the house calling the Star Prairie sheriff saying, "I just killed my wife. No, I didn't—here she just walked through the door!" The sheriff come out. I was scared to death. Blood was coming out of my mouth and nose. The sheriff wanted to take me to [a] battered women['s shelter]. But—silly me, I did not want people to know what he was really like. I still was going to hide it.

That domestic violence can be the grim underside of economic hardship is not news. Nor is there evidence to suggest that the farm crisis was an exception in this regard.[9] Yet Jeannie's private terror and concern for secrecy involve something more, I believe, than a battered woman's desire to "protect" her abuser.[10] For Martin was not the only one who wanted her to remain on the farm. Farmers Home and Farm Credit were also interested in having their investments yield a profit. As long as the value of collateral property made the Travis loan "look good" on paper, there was little incentive to probe beneath the surface and ask whether the loan was good for the family involved. In this respect, lenders were inclined to act like polite neighbors who avoid talk of private distress until it becomes impossible to ignore. Yet Jeannie's case shows us another side of the "open secret." While she may have tried to hide her domestic abuse, she did not conceal the truth of her economic situation. Quite the contrary: she actively sought help for her financial crisis, only to find no one willing to respond.

Hog farmers Roger and Ellen Guntley could have understood her problem, had there been a way to discuss it. The Guntleys had taken over Roger's family farm in the early 1970s with a loan from the Federal Land Bank. When rising interest rates caught them up short in 1982, the Farmers Home Administration issued a second mortgage on the farm, allowing them to pay down existing debts and continue farming. But they soon felt patronized and demoralized by FmHA's accounting procedures.

ROGER: FHA wanted a certain percentage of the pig checks. You had to account to them. No matter what you sold, you had to go in there. They would sign a check, and you had to tell them what you're going to do with that check. That's where you feel like you've lost it. You had no control.

ELLEN: I started back to work to supplement the income in 1982. When you made out your projections that fall, you were to tell them how much I earned. To me, this was a great invasion into your privacy. You had to account for everything you spent. So, if you took a vacation, you'd feel guilty. If we bought a radio or something, you'd feel guilty that this should be going to FHA. This isn't *your money,* you know. And people would say to you, "Well, my goodness! With these government programs and so forth, you people can't make it? You don't belong in farming. You're just poor managers." I mean, this was very difficult.

To the Guntleys, this level of public surveillance was unendurable. In 1986 they decided they wanted to get out of farming without having to declare bankruptcy. They proceeded to request mediation through the county extension service, but to their dismay, none of their lenders seemed willing to work with them.

ROGER: The guy they had representing the Federal Land Bank had absolutely no power to do anything. He was just a wimp. We'd make proposals and he says, "Well, I don't know. I'll have to go back and talk to my superiors about that." We'd propose all kinds of different ideas, but they didn't do anything.

ELLEN: The guy from FHA would never come. He was afraid of being shot. There was a time in which, seriously, he was afraid that somebody was going to shoot him. So he didn't come to mediation meetings.

ROGER: A farmer tried to run him over with a tractor. I remember one time he was in his office and a car backfired and he ducked behind the desk!

ELLEN: Because they were being shot at the time. People were angry. We weren't angry. We just wanted a way out.

It is perhaps one of the lasting ironies of the farm crisis that it was virtually impossible to *voluntarily* leave farming by way of the "lender of last resort." The agency that was designed to help farmers stay on the farm was also often the agency that actively discouraged farmers from taking control of their situation when the writing was on the wall. In the early 1980s, FmHA county supervisors were under considerable pressure to move federal dollars into the farm sector, with the goal of alleviating economic distress and rural discontent. In this political climate, loan officers were instructed *not* to turn down farm loans if they could be made to "look good" on paper. Bob Skoglund describes what it was like to work at Farmers Home after President Ronald Reagan took office in 1980.

BOB: [Reagan's] people very firmly believed that their economic program was going to be so dynamic that we couldn't make a bad loan. I remember our state director when he was still newly appointed coming around to gather a bunch of us supervisors together and saying, "You *make* these plans work. Use whatever prices you have to, because by fall, the economic program will be working so good that all these will be good loans, you just wait and see." So we did.

KATE: When he said "make the plans work," what did he mean?

BOB: The budgets, the cash flows, whatever it takes. Put in the price of corn higher than you ever expect it to be.

KATE: How did you feel about carrying out those directives?

BOB: Well, it probably took me a year or two before I could see that, hey, year after year, these farmers don't pay their debts—and they borrow more money to cover their last year's bills, and it can't go on. It's not sound lending. But it wasn't popular to turn down loans, so you just made the plan work the next year. Even though, if anybody asked you [if the numbers were fudged], you'd say, "Heck, it didn't happen!" On paper, it worked. Then, in a year, you'd refinance the unpaid bills and refinance the unpaid payments, and do it again. When land went up, you could always lend the guy more money.

Pumping money into rural America may have been politically expedient, but the implementation of federal policy collided with a community culture that fatally amplified the institutionalized "ignorance" of Farmers Home. When county supervisors and loan officers were authorized to examine the intimate details of a farm family's life—but took what they saw to create an artificial portrait of financial stability—they reproduced, in a particularly virulent form, the social structure of the "open secret." For there was nothing more painful to farmers in distress than the abject humiliation of living with the fact that their financial problems were "known" to all, but unspeakable in public and rendered invisible on paper.

WILLY, *with wonder* I was driving along, you understand? And I was fine. I was even observing the scenery. You can imagine, me looking at scenery, on the road every week of my life. But it's so beautiful up there, Linda, the trees are so thick, and the sun is warm. I opened the windshield and just let the warm air bathe over me. And then all of a sudden I'm goin' off the road! I'm tellin' ya, I absolutely forgot I was driving. If I'd've gone the other way over the white line I might've killed somebody. So I went on again—and five minutes later I'm dreamin' again, and I nearly—*He presses two fingers against his eyes.* I have such thoughts, I have such strange thoughts.

—Arthur Miller, *Death of a Salesman*

7 Social Trauma

When Luverne Dahl talks about losing his dairy farm, he gets a faraway look in his eyes. In 1981, on the basis of land valued at $1,000 an acre, he and his wife, Betty, upgraded their dairy operation with a loan from the Production Credit Association. There would have been no problem, Luverne insists, had the bottom not dropped out of the real estate market in 1983. Sitting on land that had lost half its former value, he could not come up with additional collateral to secure his loan. Production Credit moved to foreclosure and held a public auction in 1985. Able to recoup only a fraction of what the Dahls' farm had once been worth, PCA charged off $10,000 of their outstanding debt.

In the best of times, an auction is the penultimate moment when, in anticipation of a secure retirement, farmers lay before the community's gaze all they have accomplished in their lifetime. Traditionally, it is only when the estate is assessed or the farm is sold that the cash value of a farm is realized—hence the adage, "a farmer isn't worth anything until he's dead." But the deluge of forced sales during the farm crisis wrought havoc with this day of reckoning. What might otherwise have been a dignified end to a respectable career was—in a contracting farm economy—an utterly debasing experience. Where farmers might once have arrayed their worldly goods before grateful heirs and appreciative neighbors, they now found themselves standing before the community's judgment only to be found wanting.

Luverne says aloud what few find the words for:

There were farmers, they took this as a sign of failure. And then the old Norwegian—you know, a man don't cry. Well, some of them couldn't cry. They ended up down in the barn and hung themselves. That's just the plain old gospel truth. Those were good men, a lot of 'em. Good family men. Good upstanding character. But it just got to 'em so they couldn't stand another minute of it. And you know enough about depression, when you get into there, you say, "OK, it'll just be better off for everybody else if I'm gone." That's distorted thinking but, then, I also know enough about depression that when you get to a certain point in there, you're not rational anymore. Whatever person it is, when they cross that one line, they are not responsible for their actions, whatever it may be—attempted suicide, if they make it, or run away, or whatever.

As I listen to Luverne talk, I have no doubt that he was once at that "point in there," or very close to it. Today he has reconciled himself to the demands of an assembly line and commutes over a hundred miles to the factory each day. "It's been a learning experience," he says of this work, frowning as he looks out the window of his small bungalow in town. "But a farmer is meant to be outside."

The Auction

"It was a very hard day," Kathy Schroeder says. "It's hard to see people come in and . . . " Abruptly overcome with emotion, she presses her lips together and looks at Steve, unable to complete her sentence.

"Take your lifetime's work away," he adds with feeling. "It's just terrible what it does."

A fly surveys the remains of our lunch: sloppy joes made with home-ground beef, coleslaw, and a large bowl of potato chips. The Schroeders' youngest child licks a Popsicle avidly, watching my tape recorder as though it were alive. "Can you talk about it?" I ask.

KATHY: We let our kids stay home from school that day. But when the auction was going on, our son was out there watching, and when they were bidding on some new equipment, he was getting real upset. He came over by me crying and said, "The tractor's gone!" I knew the tractor didn't sell and I told him that.

STEVE: See, the whole psychology shifted to the negative by that time. Nobody would pay nothing for anything. So, when you had an auction, you were not only taking a low price, you were taking a *discounted* price.

KATHY: Everything was going, too. We got taken in every way. It's like [people thought], How much can we *beat* on these people?

STEVE: They were taking every damn thing we got.

KATHY: Everyone was pounding on us.

In a deflationary economy, a forced sale can feel like a public flogging.[1] There is a powerful sense of being taken twice—first by the global conditions that make debt impossible to repay, and then by the local conditions that make farm property impossible to resell. If the global market inflicts the initial blow, then it is the local market that adds insult to injury. Moreover, where global forces seem to operate in invisible and anonymous ways, local forces have a human face—that man in the pork association cap who is bidding on your tractor, that neighbor who has been eyeing your land for years. It is at the local level where ritualized forms of economic dislocation play out—seizing upon regional customs and styles, and populating the scene with regular folks, some of whom you have known all your life.

An auction gives ritual form to the experience of farm loss. Even for farmers who never go through one—either because they avert foreclosure or their lender decides against immediate liquidation—it is the auction that structures the cultural meaning of dispossession. Central to the ritual process is a confrontation between buyer and seller. The interests of these parties are at cross-purposes from the outset: the buyer is looking for a good deal and the seller wants to get a good price. Only by submitting the goods in question to the supply-and-demand calculus of the market can a "fair price" be established—and, in an important sense, legitimized. When all is said and done, making the transition from "farmer" to "ex-farmer" is a matter of transferring the ownership of productive assets from one individual or family to another. The *transfer of ownership* is, therefore, the social function of the ritual. As with all rituals, the community's underlying concern is to see that individuals make the move from one status to the next without a hitch. Should a wayward soul become suspended between states—in what is called the "liminal" phase of the ritual process—the social order of the community would be seriously threatened.[2] There is no place in Star Prairie's cosmology for someone who is "betwixt and between" ownership and dispossession—neither fully absolved of, nor legally accountable for, outstanding debt. It is the auction that forces movement between these social states.

What may go smoothly in voluntary farm exits rarely does so in a contested foreclosure. Here, as we might imagine, there can be considerable resistance to the ritual process. Even when a farmer does not take up battle against lenders through lawyers or a tractorcade, it is excruciatingly painful to accept the social identity of one who has "failed." This is why, for so many of Star Prairie's farmers, the experience of farm loss had a traumatic quality. As Steve and Kathy Schroeder attest, the ritual process is *punitive.* It felt like people were beating and pounding on them—trying to take as much as they could while paying as little as possible. When the crowd dispersed and a valuable tractor was left standing there, unsold, it became a sign, for Steve and his son, of just how mean and senseless this struggle was:

I'll never forget it. My boy, that night, he bawled his guts out and I did too. We bawled our hearts out. He went out there and he says,

"Dad, is the tractor *really* ours?" It's the hardest thing you'll ever do. Explain that to your kids. I just told him, "Yeah, it didn't sell." It was unbelievable that it didn't sell. It's just kind of funny. He had the hardest time. [*His voice breaks.*] And, through it all, what it boils down to is your love for farming. Farming becomes a love. And the reason you love it is because you feel part of it. You can see what you're doing when you're accomplishing it. Our boy, he wanted to farm in the worst way. What happens is, you figure, well, if I'm not getting paid for my commodities, that's all right, as long as I can make the payments and there's a future here—that my family can farm and make a living at. It'll get better and it'll be all right. That's always your hope. What happens is, after you've had that hope dashed and you're down to nothing—and your family looks at you and says, "You know, you're a real failure"—then it really hits home. Your boy looks at you and says, "Dad, they can't take the tractor!" Because he could drive it, he could operate it. It really hurts, big-time.

The auction that divests farmers of ownership operates to ensure that they are "down to nothing" in social as well as material terms. More than things disappear. With them goes a whole means of livelihood, for you can't be a farmer without a tractor. But to stand revealed as a "failure" before the community is not the only, nor even the most painful, blow inflicted by the ritual process. Where farm loss "really hits home," as Steve acknowledges, is *at home* in the private sphere of the family. Because farming is an intergenerational project, the involuntary transfer of farm property to anyone other than designated heirs strikes at the heart of what a family farm *is*. Failure, in this context, spans the fullness of time, as a legacy is lost and the birthright of future generations is foreclosed. The prospect of being exposed as the weak link in this chain was more than some could bear.

"Some farmers committed suicide over this situation," Patricia Alber says. "Because men have egos—like it or not—and if a farm has failed, the whole thing reflects on you personally." Her husband, Roland, nods and looks a little sheepish.

PAT: Now, Roland here got really sick. He was in the hospital a couple days.

ROLAND: It was a stress-related factor.

PAT: At the beginning we didn't realize what it was.

ROLAND: I didn't even know what was happening. Didn't think it was even happening. I would go to the field, which to me is a real relaxing way to spend the day, and sit on the tractor and cultivate corn. I'll tell you what, I could make about two rounds and I'd come apart. Like you come apart involuntarily. You don't know why it's happening. You can roll on the ground all over the place, just like somebody hit you with a great big slug, and you can't stop it. No way.

PAT: And we weren't being that closed-mouth about our situation. We've always communicated, and we have family we communicate with. But you're walking around a family farm like this, and you feel like all the ancestors are there watching everything you do, from generations back. And if you fail, you're failing *all* the family. Which is ludicrous really, but these are the feelings. I know there are a lot of other farm families that went through the same thing, but they would never talk about anything like that.

At the moment Roland falls off his tractor in physical pain, the Alber farm is not on the auction block. Yet the feelings he describes sound uncannily like those reported by the Schroeders: in the course of doing what he finds most relaxing, he feels himself "coming apart involuntarily," as if he has been hit with a "great big slug." Once again, although the community's participation in the ritual process is crucial, it is at home, in the heart, that farm loss hurts. Instead of looking into the eyes of a tormented son, Roland feels the eyes of his ancestors upon him, watching his every move. The acute physical pain he feels may be "stress related," if by that we mean, as medical practitioners usually do, a high level of physiological arousal caused by social or environmental conditions.[3] However, if we want to understand what this pain means—indeed, if we want to know what social conditions *cause it*—then medical science will not get us very far. For the pain of farm loss is a cultural phenomenon. It is part and parcel of a culture in which members of the community are empowered to ignore—and profit from—the suffering of their neighbor, even when that suffering is no secret and right next door.

Scenes of Suffering

For Jane Gunderson, Star Prairie has always been a special place. Her great-grandparents arrived with the first wave of settlers after the Civil War, and today she has many relatives still living in the area. "There's a closeness here that you can't find anywhere else," she explains. "You walk down the street and everyone knows you, everyone greets you. People ask you how you *are,* and they're interested and concerned. It's just home. To me, it'll always be that way." After an unhappy first marriage took her to Minneapolis, she come home to Star Prairie as a single mother with three children. Harold Gunderson, her old high school flame, had remained a bachelor all these years—just "waiting for her," he says now, with an affectionate smile. Their marriage in 1976, as Jane recalls, was like a dream come true:

We were active in the community. He was on several boards in town, president of the congregation at church, and I was involved in all the school activities. We were optimistic about everything. We didn't dress elaborately, but we had nice clothing and a nice vehicle. Our home was old, but it was very well kept. We were just happy. It didn't seem like it would end, you know? It was just like a dream. To me.

Harold's ties to the community, while significant, were not as deep or long-standing as hers. His parents rented a farm in southern Minnesota until they moved to Star Prairie in the 1950s, able to purchase a farm of their own with help from the Farmers Home Administration. Harold, the only son, began farming at the age most farmers do: when they are big enough to drive a tractor. By the mid-1960s, having worked side by side with his aging father for many years, he was able to take over the farm when he turned twenty-one, again with assistance from Farmers Home. During the 1970s, hard work coupled with a booming farm economy won him considerable recognition, including the county's own Young Farmer of the Year award.

The Gundersons had gone into debt in the late 1970s with the support of Farmers Home and the Production Credit Association. They invested in machinery and storage bins and built a new house on the assumption that worldwide demand for American grain would remain strong. But their optimism, Harold angrily recalls, was short-lived:

> Jimmy Carter's grain embargo just devastated us and lots of other people. It destroyed our crop prices. We had a real good thing going. We were going into the late summer with the prospect of excellent prices. When he put the grain embargo on and shut the pipeline off to Russia, everything became worth nothing. So then, by the following spring, interest rates were high. Every year it got a little worse. You had to generate more money to make everything work. But it hadn't been a problem up until then.

By 1983, with poor crop receipts for two years running, the Gundersons were unable to pay off their operating loan with Production Credit. Harold had received assurances that he would receive a loan for spring planting and that the previous years' debt would be carried over and reamortized. But when he went into the agency to complete the paperwork for this loan, he was told that it had been denied. Too much risk was involved, his loan officer informed him, and PCA was "pulling in its horns." With several thousand dollars of seed, fertilizer, and chemical supplies already charged at the local co-op, the Gundersons felt they had no choice but to mortgage the house they had purchased for Harold's parents in town and borrow the money they needed from the Star Prairie Bank. As it happened, securing a loan to put in that spring's crop was the least of their worries.

Arranging the day's routine around the mail delivery, Harold managed to keep correspondence from lenders to himself. Wishing to shelter Jane and the children from what was happening, he struggled to create the impression that although money was tight, no drastic changes in their lifestyle were necessary. When Jane hesitated to buy a new winter coat, he encouraged her to go ahead, even if it meant cutting corners elsewhere. When their daughter needed tuition money to take classes at the local technical college, he dug deep in his pockets to help her, never letting on that he was "borrowing from Peter to pay Paul." So focused was he on keeping the farm together that when his father died, he was unprepared for the emotional impact of this loss. Although the elder Gunderson's illness had been diagnosed several years earlier, his death ushered in an overwhelming sense of futility. Burying himself in work from morning to night, Harold harvested his crops and paid creditors what he could. His only hope for financial recovery lay in getting a good price for the soybeans he was storing for sale later that winter. When the day came to market them, however, he discovered that the whole bin had soured.

JANE: It was worth many thousands of dollars. That was a kick in the head that you didn't count on until it happened.

HAROLD: Yeah, I tell ya, I didn't think there was any problem with the beans. I stuck the auger in the bin and I backed the truck up and I started the auger, and this crap started to come out and I knew that that was it. I knew this is the straw that'll break the camel's back. I tell ya, I laid down on the ground and cried. For whatever good it did.

While Jane realized the seriousness of the loss, she did not realize its impact on Harold until she woke up one morning shortly thereafter and found him curled up in a fetal position beside her, unable to respond. Over a decade later, she admits, "I've never been so scared in my life."

Harold was hospitalized for a week and released with a prescription tranquilizer. But he soon found it difficult to manage the farm under sedation. "These pills were supposed to relax you," he complains. "Well, I was *relaxed*, I tell ya. I didn't give a damn about anything." Frustrated with treatment that only seemed to compound his despair, he stopped taking his medication and refused to continue therapy with the counselor their pastor had recommended, because, as he puts it, "she didn't know a lot about stress." Increasingly, he turned to God, finding in this more deeply committed spirituality the solace he was seeking.

HAROLD: Sometimes God just has to just knock you down on your knees and get your attention. It's probably all for the best. And even if it isn't, you might as well look at it that way and retain some of your optimism. But it certainly has been a learning experience.

KATE: What did you learn?

HAROLD: Well, I'm so darned stupid, you see. It takes awhile before you get to the point where you know there's nothing left that you can do, so you have to just give the whole thing up to God and let him work it out for ya. Indeed, He has. It isn't exactly the way *I* would have planned it. But we're still here.

If the farm crisis brought Harold closer to God, it was no thanks to their Lutheran church. Their pastor, they both agree, was "uninspiring," but even more to their dismay, members of their congregation began to treat them as though they "had the plague." Where one would have hoped for compassion and offers of moral support, there were cold shoulders and averted eyes. Jane, more than Harold, found this ostracism especially painful, as it ran counter to everything she believed Star Prairie should be:

You were really alone. We really felt alone. I was surprised. I mean, we were respectable. We hadn't done anything. We hadn't robbed anybody or cheated anybody. We ran into financial trouble, and the community just backed off. They didn't want to get into it. They didn't want to inquire as to what had happened, because they might be called on to respond, and nobody wanted to do that. They just wanted to keep their distance. And that was very hurtful. Even today—sometimes I run into a person and I think, Well, you were in my circle at church—why didn't you come and help me?

Jane insists that she was not expecting economic assistance or a dramatic display of solidarity. Saying hello and asking her how she was would have been sufficient. Inviting her over for a cup of coffee would have been truly appreciated. Years before, when she returned to Star Prairie as a divorcée, the community had welcomed her with open arms. But when it came to losing the farm, she says, the same people seemed remote and unforgiving. Whatever the community's reasons for withholding support, the effect is tremendously punishing. Like criminals or outcasts, distressed farmers are made to feel they possess a deeply discrediting character flaw. Neighbors may try to ignore this blemish out of "politeness," but like a scarlet letter, it is always visible and always potentially mortifying.[4] In essence, the community says to farm losers: "We will act as if we don't know you are in trouble, as long as you act as if you are not in trouble." Far from demonstrating "respect" for people's privacy, the norms of rural neighborliness place the onus of *hiding* financial hardship on those who are *already known* to possess this stigma.

As Jane recalls, the most painful part of their whole ordeal was trying to "hide" what others already "knew."

JANE: I remember going to the grocery store one time, and I can still see myself standing there in my red winter coat. I had to buy groceries, and I had *X* amount of dollars to do it with. I had tried to total it up in the basket when I was going down the aisles so I wouldn't go over. I was under by two pennies. I thought my heart was going to stop, standing right there, you know, watching that till ring up. I thought, Oh no, maybe I shouldn't have bought that. Maybe the cats could eat bread and water or something. It was the most horrible feeling. I'll never get over that *feeling* I had.

KATE: What was the feeling? What if you had been a little short?

JANE: Well, I would've had to take something back and put it on the shelf, and everyone was in line waiting. And everyone knew we were in trouble, and it would've been just *awful*. So I was just relieved when I got my two pennies back.

The local supermarket provides a stage for the ritual drama of farm loss in a way that distant commodity markets do not. Like an auction in one's own backyard, local scenes put human faces on the economic forces that constitute capitalist society. While financial crisis may originate in global grain markets or domestic monetary policy, the cultural meaning of that crisis is determined at the local level. Who among us has not had to rummage for change in a checkout line or jettison an item or two? Only under circumstances that strike at the core of a community's sense of moral order does a scene like the one Jane describes take on epic proportions. Before the judgment of the cash register and the watchful eyes of her peers, her consumption choices are under scrutiny every bit as much as they would be in a foreclosure auction. Is she trying to live beyond her means? Is she dressed too elaborately? Is cat food a necessity or a luxury? Indeed, what kind of dignity can she ultimately preserve, if it can be lost for want of a mere two cents?

Farm losers in Star Prairie recognize that the source of their distress is not local. They know that there are larger forces at work and that the central drama of their lives cannot be reduced to the vaudeville of capitalist villains and rural innocents. It is not their neighbor, nor even their lender, that they blame for their distress. Yet to the extent that they are able to blame anything at all, it is the local scene and its culturally choreographed "ignorance." For if their original loss was caused by the invisible hand of the market, then local scenes of suffering realize that loss, clarifying with every reenactment its inescapably social consequences. Even Harold Gunderson, who reels from a blow of divine origin—"sometimes God just has to knock you down to get your attention"—describes a scene in which it is not God's judgment he fears, but that of his peers.[5] When he drops to the ground in front of his bin of sour beans, it is knowing what other farmers will think that reduces him to tears. For his spoiled beans make visible—to himself and others—what the community already knows. Like Jane's red coat, they represent the awareness of failure: they are a sign of the stigma he cannot hide.

Social Trauma

Doctor Ludke had been the Berg's family practitioner for over fifteen years. Peggy was particularly fond of Marge, his receptionist, who always tried to spare Peggy the hour-long drive to the clinic, if at all possible. Listening to descriptions of an illness or injury over the phone, Marge could be relied upon to distinguish problems that required medical attention from those that did not. When the children had come down with strep throat, one after the other, she saved everyone the trouble of repeated office visits by phoning in prescriptions at the local pharmacy. Because they seemed to understand each other, as women and as mothers, emergencies were that much less stressful. Yet in the weeks following her husband Dewayne's disappearance, Peggy had reason to doubt the strength of their friendship.

Advised by a farm advocate to get her legal affairs in order, Peggy went to the bank to collect the necessary documents. Opening the safe-deposit box, she was stunned to discover that several personal items—a watch, a child's poem, a family photograph—were missing. Taking this as a sign that Dewayne was still alive, she checked to see when he had last been there. The bank's register indicated that he had been in the vault the week before he vanished. Peggy left the bank in a daze:

I walked out to the parking lot. I have a full-sized Chevy van, but I couldn't find it. *Big* parking lot—probably has fifty parking spots. So I sat down on the curb, on the planter, with my bags of papers and waited until the bank closed and these people left. And there it was! It was in the lot all the while. I couldn't find it. So I drove to my family doctor.

When she arrived at the clinic, her experience of acute distress was met with a lack of comprehension that bordered on rudeness:

I got there and I said, "I'm here to see Doctor Ludke," and Marge says, "What's wrong?" And I said, "I just have to see Doctor Ludke." I couldn't figure out how she had gotten so stupid. I mean, couldn't she *see* that I had lost my van and it wasn't lost? There *must* be something wrong with me. So she came back to me and she said, "Peggy, are you having headaches?" "No." "OK, what do you want me to tell Doctor Ludke?" "I just need to see Doctor Ludke." So she walked back again and then she comes in [and asks], "Are you having menstrual problems or stomachaches?" "No," and I'm thinking, Marge, can't you see? I've just lost my van, obviously something's wrong with me. And my chest hurts a whole lot, Marge, but when did you get so dumb?

After the clinic closed and everyone else was gone, Doctor Ludke came out to the reception area to talk with her:

He said, "What you have I can't help you with." I said, "But it hurts so bad here [*indicating her chest*]." He said, "There's no medicine I can give you, Peggy, but I must tell you that I'm going to be hospitalizing you if these symptoms continue." I said, "*I lost my van.* I don't know what's going on. I sat on a curb and now I'm here." He says, "You must take a day off." "OK." "Yes," he said. "Every week you must take a day off and you must leave the farm."

Even in the retelling, Peggy seems astonished to realize that her trusted receptionist and doctor seemed incapable of responding to her distress. That they did not take her "lost van" seriously—we can imagine the "knowing" winks exchanged behind her back—only served to compound the pain she felt. Moreover, Doctor Ludke's advice did little to help the situation. Attempting to follow his orders and "take a day off," Peggy found that trying to "relax" only made her symptoms worse:

[The boys and I] went to the state park in South Dakota and had our picnic lunch and I tried to read. I'd read a page, and I realized, This page made no sense. So I tried again. Then I tried to make little notes on the side to remind myself what I read—and I'm talking light literature, OK?—and it still does not hang together. *I do not read now.* I'm thinking, This is bad when you can't read.

Like Roland Alber, Peggy feels herself "coming apart involuntarily." Instead of falling to the ground in physical pain, it is the inability to concentrate that incapacitates her. And like

Steve Schroeder, she cannot explain to someone else what it means—and feels like—to lose something that is *not actually lost.* Much as Steve's tractor reappears at the end of the auction—not sold, but not fully belonging to him either—Peggy's van shows up in the parking lot—not lost, but not fully "there" anymore, either. Caught up in the ritual process of dispossession, the tractor and the van have been alienated from their owners, yet remain suspended in a kind of limbo, belonging fully to no one. That Peggy's "lost van" is anticipated, not actual, does not alter the fact that, in her mind, the ritual process has already begun. No creditor's notice has arrived, since the Bergs were not, as yet, technically in default. But at the bank that day, Peggy is forced to acknowledge that Dewayne's disappearance was premeditated. In the privacy of the vault, he "appears" to her in the missing personal mementos—revealing himself to be not actually *missing,* but not fully present, either.

The trauma of farm loss lies in the recurrent experience of a specific kind of social scene. If the ritual process involves the forced movement from ownership to dispossession, then at every point along the way, distressed farmers are subjected to social situations that reenact the moral drama of this transition. At each point, farm losers are confronted with a sign of debt and dispossession—of what they owe and must repay—and, in each case, they are made to feel that the world is slipping away from them and that they have done something to deserve it. "Social trauma," as philosopher Judith Butler observes, "takes the form . . . of an ongoing subjugation, the restaging of injury through signs that occlude and reenact the scene."[6] Farm loss is a social—not naturally induced—trauma. The erosion of human dignity that accompanies it happens at the hands of those who are your friends. For it is the behavior of your neighbors, Jane and Harold Gunderson observe, that makes economic loss feel as though someone has died.

JANE: Another thing that bothered me—all our neighbors were just like flies to honey. They all wanted the land. You know, they could hardly wait till it was all over so that they could pile in and buy the land. We *needed* a buyer, but it was yet, it was, to me—it was like they couldn't wait for the horse to die so they could pick at it.

HAROLD: It's the same thing when a farmer dies—there's somebody there to rent the land before the body's cold, and I'm not kiddin' ya. I mean, it is that—it's that tough a game.

Like many farmers who lost their farms after the passage of Minnesota's 1986 homestead protection law, the Gundersons and the Schroeders were able to repurchase their house and homesite after their foreclosure.[7] Yet the opportunity to remain in their own homes came with a stiff price: for as long as they stay in Star Prairie, they will be subjected to the sight of watching someone else farm their land. "I could've killed him," Steve says, referring to the neighbor who jumped in and bought his land—paying more than its current value in cash—before the Schroeders had a chance to work out a rental agreement with the insurance company.

STEVE: I can understand how people can literally—you know, there were bankers that were shot—I can understand that.

KATE: What makes people do something like that?

KATHY: It's the hurt that this farm is what we've been working for, and it's gone, and now he's taking over. It's just like he's trying to take everything you have away from you. And it's like he doesn't belong there.

STEVE: And that's the same thing you feel with the banker. You feel like, Look, I sweated and I worked for this son of a gun, and I *deserve it.*

Farm loss involves a ritual process that produces an ongoing social trauma. If farm loss is like a death—or makes you want to inflict one—it is a socially constructed kind of death, one that comes from having "everything you have" forcibly taken away. The Schroeders know that there is a profound difference between the death of a loved one and the loss of a farm. At the age of ten, their oldest son, Troy, was eager to assist them with every farm chore, no matter how strenuous or what the hour. One night in 1982, while he was helping Kathy unload a truck full of soybeans, the bean-filled gravity box became off-centered and suddenly flipped over. Troy was struck down and pinned beneath it. The box was too heavy for Kathy to move alone, and despite her immediate call for help over the farm radio, by the time Steve could make it in from the field, Troy had suffocated. When the foreclosure notice came the next year, Kathy recalls, "it didn't even really hit us—we were just, you know, still in a daze."

Steve agrees. "We had the greater loss first," he says. "But it was a double loss, because it was hard for us to accept the farm *and* the son. We were so heartbroken that we were walking around in a daze." Through this period, there was concern about Steve being suicidal. In the end, he says, it was faith in God's divine plan that gave him the will to go on. Yet the Schroeders continue to feel alone in their grief.

KATHY: Even fellow farmers can't appreciate it. There's so many things you cannot appreciate, and no matter how much you tell someone, they cannot appreciate it until they've gone through it themselves.

KATE: Like what?

KATHY: What it's like to lose someone you love. What it's like to lose your farm. What it's like to watch your machinery be auctioned off. What it's like to see your family in pain. You can't appreciate that unless it really hits *you*. I have known some friends, they lost someone, and I think about it. Well, I try to think about, how do they feel? You know? And it's like, Oh gosh, that hurts too much, no, I'm not going to think about that. Until you're forced to live it, you have no idea what those people are going through.

While it may be true that there are limits to our ability to empathize with someone else's pain, it does not follow that we have no way of "knowing" how others feel. Our knowledge of subjective realities has a social function: the more we can understand other people's pain, the greater our sense of community. Thus, it is instructive to consider why Kathy believes empathy fails: if a particular experience "hurts too much," people can choose not to "think" about it. In Star Prairie, it is clear, there are powerful reasons why people might choose not to think about the subjective experience of farm loss. Indeed, were it not for cultural limitations on empathy, ritual dispossession would grind to an abrupt halt. Who would buy repossessed land, livestock, and machinery, if not those who "pile in" and have few qualms about picking the carcass clean? For who, in practice, sees the ritual through to its bitter end, if not the friends and neighbors whose compassion is in short supply?

"Lou and Oscar can't see those things," said Alexandra suddenly. "Suppose I do will my land to their children, what difference will that make? The land belongs to the future, Carl; that's the way it seems to me. How many of the names on the county clerk's plat will be there in fifty years? I might as well try to will the sunset over there to my brother's children. We come and go, but the land is always here. And the people who love it and understand it are the people who own it—for a little while."

—Willa Cather, *O Pioneers!*

8 The Last Farmer

Most of Star Prairie's farmers tell me that they are not in farming for the money. If they were, they laugh, it would be a bitter disappointment. Instead, they value the spirit of independent enterprise and individual initiative that owning their own business demands. For many, farming has value as a way of life precisely because it creates economic opportunity for hard-working families and offers a form of "job security" that is increasingly hard to find in America today. However precarious the independence or compromised the security, family farming presents itself to those who undertake it as a decisive alternative to factory or service work. The farm—as a business and as a residence—brings together two aspects of class identity: the "productive" identity we derive from the kind of work we do, and the "lifestyle" identity we demonstrate as a result of the consumer choices we make. In contrast to well-paid blue-collar workers, who often identify as "working" men and women on the job but "middle" class at home,[1] farmers generally identify as "middle" class, even though their household income and the material lifestyle it supports would, by conventional measures, be considered "working" class. Farmers claim to be middle class because being their "own boss" allows them to enjoy a lifestyle that money alone cannot buy. As Virgil Thompson observes, this class identity depends as much on what one "does," as on what one "does without":

Agriculture is—it's not hard to make a living, but it's hard to [live] within your capability of earning, your earning power. You have to be satisfied. That was why I wanted to do the interview on the lawn over there, so that you could sit out there and see the beauty. I was hoping that it would be a beautiful sunshiny day and you could see one of the reasons why I'm out here. There's pheasants in the area and you can hear their call, and once you get acquainted with it, I can pick out three different families of pheasants in different parts of the farm. That's why I stay out here in agriculture. If you're out here because you think you're going to get rich, you can just as well go work for somebody because it's not going to happen. But wealth is in the sound of the pheasants and the seasons changing and all the things that go with it. That's how I feel about it.

The afternoon I spent with Virgil was overcast and stormy. Rain beat down steadily on the roof of his old farmhouse, and several times during our conversation the electricity flickered off, then on again, as loud claps of thunder punctuated our words. Yet it wasn't difficult to imagine what he had hoped I would see. His farm is not far from where my great-grandfather began farming in 1920. The gently rolling land in this part of Star Prairie is tufted with dense stands of deciduous trees and troughed with small lakes and sloughs (pronounced *slews*). Unlike the Red River Valley to the west, where the land is checkerboard-flat for as far as the eye can see, farms here are nestled into the landscape like carefully placed eggs. That there is beauty in the land and the wildlife it supports is undeniable. But what does it mean to claim that there is "wealth" in the call of a pheasant or the change of seasons? We have seen how the code of frugality establishes normative standards of consumption and suppresses conspicuous displays of wealth. But what does it take to be satisfied with the narrow profit margin endemic to agriculture? Or to put it another way, why are expressions of discontent culturally associated with failure?

Entrepreneurial Spirit

Farming, Star Prairie's farmers say, is "in our blood."[2] It is what they grew up with, and it is what they know. In an important sense, they believe, it is *in* them as part of their kinship with parents and pioneering ancestors. How do you get dirt in your blood? To hear Virgil Thompson tell it, you acquire it at an early age:

> You get dirt in your blood. My father let me drive the tractor when I was four years old, and I just loved that. It was hard work, but he made it fun. I was Dad's penny-a-day man. I would work for a penny a day, and every Saturday night I'd get a nickel, go downtown, and buy a Fudgsicle—and that was the high point of the week, you know? He taught you that if you worked, you got reward. And most of the time [in farming], there was nothing to it, you could get a good reward.

Learning to farm, Virgil insists, is a lesson in the basic principles of a capitalist society. Not only is the market figured as inherently just—"If you work, you get reward"—individual moral character is built by internalizing its logic. More than anything else, it is this inner-directed drive and ambition that makes being a farmer a rewarding occupation, although by no means an easy one. The temptation to let others do your thinking for you is ever present, and if you have not learned how to manage your time and money wisely, you will not be a successful farmer. When Virgil thinks about farmers who failed during the farm crisis, he blames widespread changes in American culture. The "conservative" management style that has stood the Thompson brothers in good stead no longer appears to be "taught" in school or learned at an elder's knee:

[My brother and I] are very, very conservative. If somebody tells you that you need a 150-horsepower tractor, go and buy a 75-horsepower tractor. If somebody wants you to have a new piece of equipment, make use of the old and do for yourself. Farming is like owning any other business, and too many people treat it like a job. When somebody tells you what to do, and you go do it, you get paid and everything is fine. But if *you've* got to figure out what to do, and then go do it, it gets complicated. *Management.* Management is the basis of the whole situation. Our schools are not producing entrepreneurs. The entrepreneurial spirit has disappeared.

Figuring out what to do and doing it on your own is, for established farmers like Virgil, the leitmotif of the "entrepreneurial spirit"—a cultural sensibility that places responsibility for the success or failure of an economic enterprise squarely on the shoulders of the individual. In the worldview of Star Prairie's farmers, the spirit of enterprise animates everything that is unique and honorable about being a farmer. A descendent of early Norwegian settlers, Virgil speaks for many in the community when he says that farmers inherited much more than land from their forebears.

VIRGIL: The individuals who took off from the East Coast [in the nineteenth century], they were looking for something. They walked out here [to the prairie] because they *wanted* to. When they got to the Mississippi River, they thought the end of the world was there, but they kept on going, you know? And that's part of the thing that is still in the rural communities and still in the farmer. Because he actually is closest to the land, and the land supports us all. He still has most of that left in him. He hears it more often from his parents, and it happens amongst us more than it does in the metropolitan areas. There's more of that in us.

KATE: More of what?

VIRGIL: More of the feeling that you *have* to do it. You *have* to go out there. You *have* to see what's on the other side. Well, if we got 125 bushels of corn an acre last year, this year we're going to get 140. Because we can do it, and the challenge is there. You go

> to work for somebody else and if you put out *X* number of units during a day, the pay is so much, but that's all you get, so that's all you do. What fueled the independent businessman is [the realization that] if I go down to my gas station in town here and open an hour earlier, maybe I'll get some extra trucks. The farmer has to go out there [and compete like that] because there are so many things that can change for him. He buys these steers and he finishes them out, but he doesn't know what he's going to get paid for 'em until he sells 'em. If he doesn't get enough to pay for his lifestyle, he's got to figure out something else. So you try a little harder. There isn't that much security offered by [working for] somebody else. [In farming], the security is in yourself and what you do yourself.

Being an entrepreneur in this community means taking pride in "doing for yourself" no matter what obstacles are tossed in your way. In particular, it means accepting the price you get for your product even—or perhaps especially—when that price is not enough to support your "lifestyle" or family living expenses. On no other point was the community at greater odds with the politics of the grassroots protest movement. Where Star Prairie's farmers attempt to find economic security in entrepreneurial individualism, activists sought to achieve it through improving government regulation and support of the agricultural economy. To local farmers at the time of the farm crisis, these appeared to be utterly incompatible pursuits.

When the Farmers Home Administration attempted to foreclose on Ray McClelland, he joined the thousands of American farmers who organized to protest low commodity-support prices by driving their tractors to Washington, D.C., in the winter of 1979. For Ray, the sight of so many farmers in economic distress was transformative:

> I began to find out that folks [in Star Prairie] are isolated, very isolated. And the FHA kept them that way, thinking, You sucker, you're the only dumb one out there that's in trouble, now why can't you pay your debt off? How come you're $2,000 behind or

$4,000 behind? You're the only one! Shit, I went out there and [saw] there's fifty thousand farmers out there from all across the United States with baseball caps on, madder than hell! I was there to learn, but *they were mad!* Well, then, I got madder and so did a lot of other people. So did a lot of farmers in Minnesota.

"Learning to get mad" was, as Ray suggests, a matter of discovering that his financial troubles were related to larger forces at work in the agricultural industry and the American economy in general. Like others drawn to the new protest movement, Ray began to study the lessons of history, looking at earlier examples of rural activism for ways to think about current conditions.[3] From the Farmers' Holiday Association of the 1930s to the National Farmers Organization (NFO) of the 1960s, he found ample precedent for collective action in the 1980s:

The farmers [in earlier movements] knew that the government was a big part of the problem because they set the price. So if that's where it's set, that's where you're going to have to go to make the change. Either you're going to have to do it [through the legislature] or you're going to have to do it the NFO way, which was set your own darn price: collective bargaining. And it's still—after all of the battles that we had; after the battle in the thirties, the battle in the sixties, and then the battle in the eighties—it's still the same damn way! It's still *somebody else* [that] sets the price for farmers. If we could sell corn on parity, we'd be getting $6 a bushel for our corn. And that's what we should be getting, because that's parity. Parity means "on par" with the rest of the economy—what you're getting or what anybody else is getting for what they're doing. Farmers aren't getting that.

The concept of parity has long been an article of faith for farm activists, past and present. In its strictest sense, it is the idea that there was once a period in agriculture—1909 to 1914, just prior to World War I—when farmers received a "fair price" for their product. In this period, it is argued, the market value of farm commodities was basically equivalent to that of industrial products. In years since, this "parity scale" has been used to determine the value of farm goods relative to the rest of the economy and other goods that farmers buy. Parity is therefore defined as "fair equivalent value between what farmers received for what they produced and what they paid for what they consumed."[4] Acreage limitation programs developed during the New Deal institutionalized the concept of parity as a way of calculating price levels that would allow farmers to maintain their standard of living over time. More commonly, however, farmers use the idea of "parity" to refer not simply to how the price of commodities should be set, but to how the work of farmers should be valued. Parity, in this sense, is a way of talking about who deserves to be middle class in America.

When Ray observes that farmers are not "on par" with the rest of the economy, he means that most farmers do not earn enough to support a middle-class standard of living. The evidence suggests that Ray is right. In 1979 farm prices reached 71 percent of parity, but by 1986 they had plunged to 51 percent—exactly the level they had been at the depth of the Great Depression.[5] In 1988 the North Central Regional Farm Survey found that 42 percent of the participating farms reported net family incomes of less than $20,000, while 36 percent reported incomes between $20,000 and $40,000.[6] Even if the latter group is considered "middle class," this is not the profile of a wealthy population. In fact, only 4 percent of those surveyed reported incomes higher than or equal to $70,000. Thus, although the value of a family's total farm assets may be relatively high—in 1988 the average was $310,972—the actual family income, after farm expenses, can be quite low. Moreover, a farmer's household income is more volatile than that of the average household, making it difficult to estimate future earnings from year to year.[7] In the view of farm activists, the only way to combat poverty and economic insecurity is to ensure that farmers receive a "fair price" for their product. Just as industrial workers must organize to secure a "living wage," they believe, farmers must organize to insist upon commodity prices that cover the cost of production and guarantee a decent standard of living.

As activists are well aware, however, higher prices are eventually passed along, in one form or another, to the consumer. During the farm crisis, therefore, considerable effort went into demonstrating public support for policy measures that might lead to higher prices at the grocery store. Central to this effort was a critique of America's "cheap food policy." David Ostendorf, former director of the Iowa-based PrairieFire, observes that a basic challenge for advocacy groups lay in making consumers realize that they were saving food dollars at the farmer's expense:

We understood that the reason farmers were going under was because they weren't getting a price, and that the prices were controlled, in effect, by government policy. So our proposal was, raise the minimum pricing for farmers, and at the same time, institute some planning measures to cut back on the massive overproduction that was driving prices down. Those were the two primary foundations of the approach we were taking. We were clear [about] saying that consumers in this country were paying the lowest prices for food in industrialized nations. We were trying to build with consumers some understanding that they had been getting low prices at the expense of farmers. And that in order [for family farms] to survive, you may have food prices go up a little bit. And we made good progress. We had a lot of polls taken during the depth of the crisis in eighty-five to eighty-six that showed a tremendous amount of popular support. Some of the polls indicated that consumers were willing to pay a little bit more if they knew that paying higher prices for food would help sustain a family farm system of agriculture.[8]

Although there is some indication that consumers are willing to pay more for their food,[9] farmers themselves remain deeply divided on this issue. As those in the protest movement were dismayed to discover, local farmers were rarely quick to endorse their call for unity and demand for higher support prices. Ray McClelland looks back on his efforts to organize farmers in Star Prairie and feels nothing but disgust for what he perceives to be their abysmal ignorance:

We could go into the [grain] elevator right now and make conversation with farmers, and the dumb shits will start arguing. "God, we need $5 corn," I'd say. And they'll say, "God, no, we don't need $5. If I had $3, I'd be just fine." But in order to be on par with everything else in our economy, we need $5. Milk needs to be $22 a hundred[weight] instead of $10. Sheep, hogs, everything—it's half of what it ought to be! But you can talk to farmers that'll argue like hell with you that that would be *too much money.* Our theory in the Ag Movement is that if you can raise prices now and keep the family farms here, we won't have corporations that will raise [prices] by twice that amount. But we couldn't convince them, so it's over. I don't think there will ever be another movement that will be so close as we were to making this happen. So the farm crisis of the eighties, it was the farmer's fault—for not organizing to set his own price. The blame goes back to the farmer.

By this reckoning, it is the family farm system of agriculture that keeps food prices as low as they are. If the trend toward larger and fewer farms continues, activists say, the day will come when giant corporations monopolize food production in the United States, and then prices *will* go up. Far from working against consumers or farmers on this issue, activists believe they are on the same side. That the protest movement seemed to be defeated by farmers themselves is, for activists like Ray, the ultimate tragedy of the farm crisis. Why was it so hard to convince farmers that they deserved a better price for their product?

Cheap Food?

The general consensus in Star Prairie is that raising price-support levels will not, in the long run, help farmers at all. Just as the wage increases received by unionized workers are eventually passed along to the consumer, it is believed, commodity price increases will eventually come back to the first and last consumer: farmers themselves. Bank vice president and part-time farmer Frank Tostrud puts it this way:

We had the NFO, the National Farmers Organization. They were trying to get a collective marketing, where we'll stop and say, "OK, you want to buy wheat or flour? We've got all the wheat here, and we're not gonna sell it to you for less than this," and thereby getting a decent price for the farmer's wheat. It's a good idea. It didn't work because farmers are quite independent people, and not enough of them wanted to, for whatever reason. But let's just suppose that it had been successful. The price of flour would have gone up, the price of bread would have gone up, all the workers producing in factories would get a wage increase under their employment contracts and the price of the factories would have gone up, the price of the cultivators would have gone up. Price of fertilizers and chemicals would go up. Then where would [farmers] be? Price fixing doesn't work. Every time they've tried that it doesn't work.

The demand for higher commodity prices, Frank argues, is basically self-defeating. The majority of Star Prairie's farmers are inclined to agree—as are most agricultural economists.[10] All are able to imagine a "bigger picture" in which higher price supports will only raise the cost of agricultural production and reduce market demand for U.S. farm products. Unlike standard economic accounts, however, local sketches of the big picture remain sympathetic to the idea that farmers are not getting a "fair price." Like Frank, farmers will often say that the NFO had a "good idea," even if, in the end, "it doesn't work." In their analysis, the problem lies not in what farmers receive for their product, but in what happens after their product leaves the farm gate. The true profiteers, they argue, are the "middlemen" who process, market, and serve up farm products to the consumer.

Tom Laslo's family has been milking ever since his father came to Star Prairie in the 1930s and built a dairy farm from scratch. Now Tom and his son, Michael, are struggling to make ends meet with a 48-cow operation, something that becomes harder to do each year. During the farm crisis, they lost a large chunk of land when they could not make payments on a contract-for-deed agreement with a neighbor. Today, they worry about generating an income large enough to support Tom's retirement and Michael's growing family. In their view, the problem is not just that farmers get a low price for their product, but that they have been receiving a progressively smaller proportion of what consumers pay for that product. Tom explains:

> See, what's wrong with agriculture is the emphasis is on cheap food. You know, they don't just come out and say that. But politically, they want a cheap grocery bill. We're told that the consumer is spending like 12 percent of his disposable income on food. Well, that's the lowest price of any country in the world. But, consequently, our share of that food dollar has dropped quite dramatically, while processing and that end of it has increased. We're down to 20 or 25 percent of the [consumer's] food dollar. Now in dairy, that's a little higher. But that has dropped. We were pretty close to 50 percent on the dairy end; well, that's down to 30 percent now. If the consumer is spending 12 to 14 percent of their disposable income, it shouldn't be going to the processor. It should be going to the guy that [produced it]. We're an important part of it. Without us there wouldn't be no processing.

As grain farmers with a small hog operation, Elford and Beatrice Hagendorf can easily find common ground with the Laslos on this issue. Not only do they share the conviction that farmers receive only a fraction of the consumer's food dollar; they emphasize the economic "gap" that exists between what they receive for their product and what they, too, must pay at the store.

EL: These [government programs] have assured our country of the cheapest food anywhere in the world by a very large margin. So anybody that bitches about the fact that we get some extra dollars better look at the grocery store because that could change dramatically if this wasn't being done.

KATE: Are you in support of that policy?

EL: No, I'm not. I'd like to see us getting our just due.

BEA: Instead of the middleman. People think the grocery prices are so high. Even in a small community like this, a farming-type community, they get the wrong impression. They think that because the price of pork chops is a dollar-whatever that we get the majority of that. And, shoot, we don't get much of a proportion of that at all! Same thing with a box of Cheerios or something. What we get for a bushel of oats doesn't even come close to this $4 box of Cheerios. They don't realize, also, that we pay that exact same price in the grocery store that they do. We're in the same boat that they're in. But we're getting such a small amount, a small price for the product. So there's this huge margin in between when it comes to the grocery store.

When it comes to the issue of what constitutes a "fair price," the Hagendorfs are firm believers in the concept of parity. The price that farmers receive for their product has not, in their view, kept pace with the rising cost of living. They feel the erosion of their buying power not only as producers faced with rising input costs, but also as consumers conscious of spending more at the checkout counter. They differ from farm activists in seeing a justification for low food prices and, thus, for the government programs that attempt to ensure them. It does no good to raise price supports, they reason, because that increase would come full circle, and the gap between their income and expenses would remain. Commodity prices are unfair, the Hagendorfs say, not because the consumer gets cheap food, but because processors, distributors, advertisers, and retailers are allowed to profit at the farmer's expense.

EL: [Back to] General Mills, the Cheerios. We're getting $1.20 for a bushel, that's thirty-two pounds of oats, and they're getting $4 for this box. And that bushel of oats could probably produce . . .

BEA: Who knows how many boxes?

EL: It's a ridiculous figure. They *steal* the product from us and turn around and retail it out and they make a bundle. Their actual cost of the grain in the cereal is probably their lowest cost. The transportation and the labor is costing them a lot more. Is that really fair? The product that it's made out of should be the most expensive. But it isn't.

Virgil Thompson makes a similar point when he says that the only way for farmers to win at this game is to control the production of their product from beginning to end:

> Sure, you can't eat oats; you want to eat cereal. But do you realize that when you give the grocer a coupon that says twenty-five cents off on this box of cereal, he gets more for handling the coupon than the guy who raised the oats? And the box costs more than the product that is the basis of the whole thing? But if you owned the box and the cereal and the whole works, you could say, "Hey, I'm not going to sell that stuff, you've got to come to me." The farmer's on the bottom of the food ladder, and he gets kicked when they go up and when they come down. Every time. But if you own the ladder, then you can control who goes up and down. That's how to be successful.

For small-scale farmers, the dream of "vertical integration" or "owning the ladder" may never be more than a gleam in the entrepreneurial eye. But it is culturally significant that when Star Prairie's farmers imagine a solution to the "farm problem," it is the lure of individual enterprise, not a call to collective action, that inspires them. Members of this community share with activists a desire to "set their own damn price," but they disagree profoundly on how to go about doing this. Rather than organizing to bargain for a fair price, they emphasize the virtues of entrepreneurial individualism and refuse to think of prices as "wages."

Potato growers Dick and Diane Porter agree that commodity programs give American consumers the cheapest food in the industrialized world. But this is not where they locate blame for their troubles, despite the fact that they experience the gap between what they earn as producers and spend as consumers with astonishing clarity.

DICK: Our food costs in this country are the lowest of any nation in the world, as far as the percentage of income that goes to food. And a lot of that money goes [to the middleman]. When we have a good potato year, we're looking at $4 a hundred. For a baked potato that's two cents apiece. I've sold potatoes eight for a penny; three years ago, we sold potatoes for twenty-five cents a hundred. [That amount] didn't quite pay for the labor to load the trucks.

DIANE: If we'd have just left them in the ground, we'd have been better off.

DICK: People don't realize where this food dollar goes. You go into a restaurant for a baked potato and I'll guarantee you on that menu it's gonna say eighty-five cents at least. One thing the government has done with these [farm programs] and keeping the family farmer out there—what they have done is keep food cheap. Because they're keeping us all out here producing. Otherwise, if it doesn't pay to produce, let the stuff lay here.

The Porters do not need to imagine the grains of oats in a box of Cheerios, the beef in a Big Mac, or the milk in a school lunch program to know that the price they receive for their product is outrageously out of proportion to the price the consumer pays. The potato not worth "taking out of the ground" will fetch a handsome price once it is on a restaurant plate, and none of that profit comes back to them. Yet this, as they see it, is no reason to leave potatoes in the dirt or drive a tractor to Washington. For not only does the demand for higher prices demonstrate an inability to manage on the subsidies that already exist; it renders you suspect of "farming the government," not the land.

Ghostly Landscapes

Dave Hurlstad recently completed a degree in agricultural economics at the University of Minnesota. After working for a short time at an investment firm in the Twin Cities, he decided to return to the family farm. "People think I'm throwing my education away," he says with an impish grin, "but deep down in, this was the place that I was meant to be." When I ask why farming appeals to him, both he and his father, George, invoke the entrepreneurial spirit.

DAVE: Everybody, I think, has the goal of being self-employed, doing something for themselves. Having the opportunity to actually *run* an organization, instead of the organization running you. There's not too many people out there that can say they have a family business that dates back almost a hundred years. This farm has always been kept within the family name. As time goes on, I think it becomes more and more of an honor for future generation to preserve that.

GEORGE: We'd never leave the community. Dave is coming back farming with us, and maybe [our youngest son] Ben. The reason is—they've worked here. Their sweat and blood is here. They've been a part of it all their lives, and they want it to continue.

Dave's decision to join the family business came with certain conditions, however. Newly engaged to a woman with a college education and career ambitions of her own, Dave made it clear to his parents that the farm would have to be enlarged to support the middle-class lifestyle that he and his fiancée aspire to. The 50-cow dairy that supported the Hurlstad family through the 1970s and 1980s simply would not generate sufficient income to support a second family—let alone a third, if Ben decides to enter the business after college. Thus, in the early 1990s, Dave and George began drafting plans for the largest dairy operation to date in Star Prairie.

GEORGE: We're undertaking a very major expansion, which is—in a sense, we'll have to admit—basically putting everything we've got in jeopardy. But we feel that's what we have to do. New generations are coming in and we have to—like I always say, you got to stay in the parade. You don't have to be a leader, but you got to stay in the parade.

DAVE: You have to be a survivor.

GEORGE: We're doing absolutely what we are thoroughly convinced that we have to do, for future generations. We're not doing it for ourselves. We're kind of breaking in a tradition. We had a 50-cow dairy operation. We are currently building up on the hill out here a 300-cow dairy operation. Our philosophy is, we have always milked in that barn that my grandfather built. We've added on to it twice. There comes a point in time, I believe, where one has to stick his neck out and jump into the next generation. So that the next generation can continue. So that we can be a profitable, viable entity.

The Hurlstads know that they are taking big risks and that they are taking them at a time when the dairy industry in the Midwest is under increasingly stiff competition from large production facilities in the western United States and Mexico. Yet when George talks about "staying in the parade" and "jumping into the next generation," it is clear that, to him, the risks of expansion pale in comparison to the risks of doing nothing to help his sons keep the farm in the family. "When I go to my grave," he says, "my hope is I can look back and say, 'Well, my farm is still going and this is what I did to help them, so they can prosper and their families can prosper.'" If this requires "putting everything in jeopardy," so be it. Family continuity was not achieved by the faint of heart in the past, he insists, and the future demands no less. Even so, George and Dave realize that others in the community do not see their project in the same light.

GEORGE: As far as community and neighbors, yes, we've taken some flak. They say we're wrecking the guy that's milking forty cows over here and forty cows over there and twenty-five cows up here. Because we're screwing it up for them.

DAVE: Flooding the market.

GEORGE: And we're doing it because of greed. We think, in a sense, that we're helping them.

DAVE: If you look at the dairy industry on a global scale, we're not competing with the guy who's two miles down the road. He's not a competitor; he's on our side. Our competitors are California, Colorado, Arizona, Florida. Mexico is a very big competitor, becoming a threat. We're trying to retain the image of the dairy industry in the Midwest, where it originally began, instead of having dairy products transported over thousands of miles and taking economic opportunity away from the people that dwell here.

The desire to cast a competitive business enterprise as a net gain for the community as a whole is hardly unique to farmers. Few of us who find ourselves in competition with coworkers or colleagues are likely to admit that our actions are motivated by selfish interests alone. Conjured up in that space between the rewards of success and the consequences of failure is an "imagined community" in which the social benefits of free enterprise outweigh the costs of competition to the individual.[11] In the name of this "community," George Hurlstad can say that he and his wife are not undertaking farm expansion "for themselves" but for the prosperity of generations to come. Although Dave imagines a different community—one comprised of all dairy farmers in the Midwest—he, too, would like to believe that what is done in the name of "survival" will also serve a greater good.

The tension between individualism and community has long been a staple of social analysis and cultural critique.[12] Time and again, thoughtful observers of the American scene have taken note of the cultural contradiction inherent in a society that aspires to realize democratic goals even as it defends the rights of individual enterprise. Often the dilemma is figured as a "conflict" between individualistic values on the one hand and communitarian values on the other.[13] But contemporary rural Americans remind us that living in a capitalist society is no easy matter of choosing between mutually exclusive values.[14] Rather, Star Prairie's farmers tell us, as the Hurlstads do, that they are "breaking in a tradition," trying to orient traditional values toward the future, even if this means "breaking with" the past in other respects. Theirs, like American culture in general, is a living tradition—responsive to technological innovation, yet deeply ambivalent about the social consequences of economic competition and the "progress" it is said to produce.

Virgil Thompson has a vision of the future. Vividly capturing the moral uncertainty inherent in the community's tradition of individual enterprise, it is a vision of the last farmer—and the ghostly landscape he will inhabit:

Before, it used to be, your neighbor a mile and a half down the road was dependent on you to help with the thrashing and silo filling and all that sort of thing. But now there's getting to be so few of us left. The way things are going now, I foresee the day when there's one farmer on the east side of the Mississippi and one on the west side. They'll be plowing and they'll meet at the river. There'll be a discussion, and shortly thereafter, there'll be one hell of a tiling project [to drain the river] and then there'll only be one farmer. And then he'll be the only one around, and he'll have trouble with his tractor on the way back, and he'll get stuck. He won't have anybody to fix it, and he'll be in worse shape than when he had a neighbor.

Virgil offers no political solution to the problem of the "last farmer." To his way of thinking, it is inevitable that this day will arrive. But when it does, he adds philosophically—when the last farmer's tractor breaks down—the land will revert to wild prairie, the frontier will be reopened for settlement, and the pioneering cycle will begin anew.

It has been a number of years since I heard Virgil's prophecy on that rainy day in Star Prairie. Yet the specter of the last farmer continues to haunt the way I think about what happened in rural America during the farm crisis and what is still happening there today.[15] Like millenarian beliefs the world over, the last farmer signifies the demise of the present social order, even as it anticipates a future in which the heroic past will return. Millenarian beliefs, and the religious movements they give rise to, are generally associated with societies that have been decimated by disease or warfare and demoralized by the failure or decline of traditional institutions.[16] Strictly speaking, farming communities have not been subjected to the same kind of violence or genocide, but they do carry the scars of chronic economic dislocation. With each family forced to leave the community due to farm failure or the lack of alternative job opportunities, a mark appears on the landscape reminding those who remain that what was once there is now gone. The sign might be as innocuous as a continuous expanse of fields where the buildings of an earlier farm once stood, or as arresting as the vacant gaze of an abandoned homestead, choked with weeds and falling to ruins. Traveling across this landscape, it is not hard to imagine what it will look like when there is only one farmer left.

Lillian Hanson thanks God that her son Luke has been able to hold on to the family farm without significant hardship. But she watches the landscape with apprehension nonetheless:

> Years ago, every little cluster of trees in the countryside here, there'd be a farm family living. And I'd like to think of them that they had a happy family farm life together. But part of this is the economy that has forced people off the farms, because you have to raise more, you have to produce more to be able to survive. The survival of the fittest. Even a small farmer has to get bigger if he's going to be able to meet the payments that are necessary. And it's a sad look in the country, really, because we're old enough to have remembered when there was a farm family on each of these places. Where there was a cluster of trees, there'd be a family living. And now that's not the case.

The "sad look in the country" that Lillian describes is, I believe, the look of a community traumatized by debt and dispossession. Trauma, in this sense, is a collective phenomena. As sociologist Kai Erikson has argued, a "collective trauma" need not have the "quality of suddenness normally associated with 'trauma,' [but] it is a form of shock all the same, a gradual realization that the community no longer exists as an effective source of support and that an important part of the self has disappeared."[17] Absorbed into the behavior and worldview of those who are suffering, a collective trauma strikes at the heart of community life, alienating neighbors from one another and filling the void with a pervasive sense of loss.

Jane and Harold Gunderson recognize that the loss of their farm is not like other kinds of loss. Their lives have gone on:

Harold is employed as an insurance salesman, and the family continues to live in the house they saved from foreclosure. But nothing will ever be the same.

JANE: With a death, you bury them and it's over. But with this, it goes on and on and on. It is that feeling that you've just lost everything that you cherished. You still have your family, and you still have your health and different things. But you have lost something that you worked hard for. And it's like something died.

KATE: What is it you worked so hard for?

HAROLD: This was our hope and our dream, the farm and the farm thing. We were farm people. It's really the only thing I've ever wanted to do. You're put on this earth for a few years and you have this wonderful work with this particular thing and you try to make it do the best you can, and therein is the satisfaction. Even if the stuff isn't worth anything when you haul it to town, when you're taking it off the field, if it's good quality and there's lots of it, you just feel good all over. It's a real high.

JANE: The land is very important. It's like your soul is in it. And you know every inch. And you know how water runs off of it, and you know what's going to grow best on that field. You have such hope every spring when you put that crop in, and, oh, it's going to be wonderful, and if it'll just rain. Every day you live with the land. Can you get the field cultivated before the rains come? Can you get the crop up before bad weather strikes? Is the wind going to blow the wheat flat? Every day you work with the land. [*Beginning to cry.*] And when it's taken away from you, it's like you lose some part of your life.

When all is said and done, the loss of a farm is not just the loss of material possessions: it is also the loss of a sense of community and one's place in the world. Farm losers feel this loss most acutely, but no member of the community is untouched by the social consequences of their misfortune. Even those who look bravely into the future know that if they fall upon hard times, they will have no one to turn to but themselves. Residents of Star Prairie understand that the Darwinian logic of capitalist enterprise puts them in competition with their own neighbors, even as they attempt to uphold a sense of community in which privacy is respected and everyone contributes to the common good. But theirs is a fragile community. The farm crisis of the 1980s dramatized, as perhaps no other disaster could, Star Prairie's fundamental inability to resist—and tacit collusion with—the forces that threaten to destroy it. In this sense, however, it is a quintessential American community.

APPENDIX: COUNTY PROFILE

The population of Star Prairie steadily declined throughout the 1980s, resting at just above 10,500 at the time of the 1990 U.S. Census. Only about 20 percent of the county's residents still live on farms, despite its predominantly rural character and continuing reliance on agriculture as a primary source of income. Unlike a growing proportion of farm-dependent regions in America today, farming in western Minnesota remains quite diversified. Thus, although individual farmers now tend to specialize in one crop or form of livestock production, the county itself supports a full range of farming operations. Dairy farming accounts for one-third of the county's total agricultural sales; and at 20 percent of the total, corn is the major crop. Cattle, hogs, and soybeans each comprise over 10 percent of the total sales, and on irrigated land, potato production has become an important specialty crop. As elsewhere in rural America, Star Prairie's declining number of farms has gone hand in hand with an increase in average farm size. About 40 percent of the farms in the county are between 180 and 500 acres, with roughly 40 percent below and 20 percent above this average. It has been within this middle category—the so-called "family farm"—that most farm loss of past decades has occurred.

By race, Star Prairie is over 99 percent white. Individuals of Norwegian ancestry comprise the largest ethnic group (about 40 percent), with those of German descent a close second (around 30 percent). Fewer than 20 percent of the county's residents have gone to college, and many (approximately 20 percent) never completed high school. The per capita income is lower than the state average, and the median household income is barely enough to raise a family of four out of poverty. A good portion of the county's low education and income levels can be attributed to the rapidly growing number of senior citizens. Over 20 percent of the county's population is 65 years of age or older, reflecting an increase of 60 percent between 1960 and 1990. Correspondingly, the average age of farm operators is also higher today than it was in the 1970s. Especially significant is the loss of farmers between the ages of 45 and 64, an age group hard hit during the 1980s. Most of the farmers interviewed for this project were in the range of 35 to 65 years old.

NOTES

Chapter One

1. Directed by Richard Pierce and starring Jessica Lange, *Country* was released in 1984. Clearly intended to be a commentary on the farm crisis, it dramatizes one family's fight to save their farm when the Farmers Home Administration moves to foreclose. The film ends with a powerful depiction of neighbors organizing to protest the auction of the family's farm equipment and household possessions. In May 1985, Jessica Lange, Sissy Spacek (*The River*), and Jane Fonda (*The Dollmaker*) drew upon their authority as "Hollywood farmwives" to urge Congress to support struggling farm families, testifying before the Agricultural Task Force of the Democratic Caucus and a special hearing of the House Agriculture Committee.

2. Not only was there no increase in political participation over this period, activists tended to be more affluent than the average farmer. Linda M. Lobao, "Organizational, Community, and Political Involvement as Responses to Rural Restructuring," in *Beyond the Amber Waves of Grain: An Examination of Social and Economic Restructuring in the Heartland,* eds. Paul Lasley, F. Larry Leistritz, Linda M. Lobao, Katherine Meyer (Boulder, Colo.: Westview Press, 1995), pp. 183–205.

3. Paul C. Rosenblatt, *Farming Is in Our Blood: Farm Families in Economic Crisis* (Ames: Iowa State University Press, 1990); and William D. Heffernan and Judith B. Heffernan, "The Farm Crisis and the Rural Community," in *New Dimensions in Rural Policy: Building upon Our Heritage*, eds. Dale Jahr, Jerry W. Johnson, and Ronald C. Wimberly (Washington, D.C.: U.S. Government Printing Office, 1986), pp. 273–80.

4. Although the establishment opinion is that modern farming methods present no significant health or environmental threat, there are those who argue cogently to the contrary. See Marty Strange, *Family Farming: A New Economic Vision* (Lincoln: University of Nebraska, 1988); and an

international report sponsored by the United Nations, *Agricultural Mechanization* (Nairobi: United Nations Environmental Programme, 1986).

5. F. Larry Leistritz and Katherine Meyer, "Farm Crisis in the Midwest: Trends and Implications," in *Beyond the Amber Waves of Grain,* Lasley et al., pp. 206–20.

6. Ibid., pp. 211–12.

7. Leo Marx, *The Machine in the Garden: Technology and the Pastoral Ideal in America* (London: Oxford University Press, 1964), p. 364.

8. See, for example, Sarah Burns, *Pastoral Inventions: Rural Life in 19th Century American Art and Culture* (Philadelphia: Temple University Press, 1989); and David C. Miller, *Dark Eden: The Swamp in 19th Century American Culture* (Cambridge: Cambridge University Press, 1989).

9. The classic analysis of conspicuous consumption is Thorstein Veblen's *Theory of the Leisure Class: An Economic Study* (New York: Macmillan, 1899). Sociologist Patrick Mooney discusses the centrality of the independent business ethos to farming in *My Own Boss?: Class Rationality and the Family Farm* (Boulder, Colo.: Westview Press, 1988).

10. In an ethnographic study of the return migration of African Americans from cities in the North to family farms in the South, anthropologist Carol Stack observes that migrants are not seeking a "promised land" of economic prosperity, nor would they find it if they were. They return instead to their birthplaces, their "proving grounds," in order to chart their progress and test their ability to survive in a place that has never ceased to represent "home." Carol Stack, *Call to Home: African Americans Reclaim the Rural South* (New York: Basic Books, 1996).

11. Farms owned by African Americans have always been among the smallest in America. In the consolidation of agriculture following the depression, their farms failed at a rate six times greater than those of whites. V. J. Banks, "Black Farmers and Their Farms," *Rural Development Research Report*, no. 59 (Washington, D.C.: Economic Research Service, U.S. Department of Agriculture, 1986); Calvin L. Beale, "The Ethnic Dimension of Persistent Poverty in Rural and Smalltown Areas," in *The Changing Situation of Rural Minorities, 1980–1990,* eds. Linda Swanson and Linda Ghelfi (Washington, D.C.: Economic Research Service, U.S. Department of Agriculture, 1995).

12. Economic Research Service, U.S. Department of Agriculture, *Understanding Rural America* (Washington, D.C.: Agricultural Information Bulletin, no. 710, February 1995), p. 5. The full passage reads: "The decline of farming employment is, in many ways, a consequence of success. Improvements in technology, crop science, and farm management have all boosted output while reducing the need for labor. Productivity growth has, in turn, led to farm consolidation, declining farm numbers, decreases in farm employment, and consequently a surplus of farm labor. Thus, the

ability to produce more with less, while benefiting many, has caused economic hardship for others."

13. For a vivid documentary of the period, see *Surviving the Dust Bowl: The American Experience*, written and produced by Chana Gazit (PBS Home Video: WGBH Educational Foundation, 1998). On the history of the dust bowl, see Mathew Paul Bonnifield, *The Dust Bowl: Men, Dirt, and Depression* (Albuquerque: University of New Mexico Press, 1979); and R. D. Hurt, *The Dust Bowl: An Agricultural and Social History* (Chicago: Nelson-Hall, 1981).

14. Taken from a family memoir written in 1991 by a Star Prairie farmer. The passage continues: "We, living on the farm, at least had enough to eat. We had meat, milk and eggs. Wheat from former crops was taken to the mill to be ground for flour. I can remember as many as 10 hundred-pound bags of flour stored in the attic of the house. My parents took out the telephone because it was a savings of $2.50 a month. That was enough to buy a pair of shoes. It was during these times [that] Franklin Roosevelt organized WPA [Works Progress Administration], where people could work on local projects to make some money."

15. Glen Elder Jr., *Children of the Great Depression* (Chicago: University of Chicago Press, 1974).

16. Postwar economists of a Keynesian persuasion did much to advance the idea that the government could and should regulate the economy. Robert J. Samuelson, *The Good Life and Its Discontents: The American Dream in the Age of Entitlement* (New York: Vintage, 1995).

17. Newman defines "symbolic dialects" as the frameworks of interpretation that filter life experiences through the historical, social, and cultural events particular to a generation. Katherine S. Newman, "Symbolic Dialects and Generations of Women: Variation in the Meaning of Post-Divorce Downward Mobility," *American Ethnologist* 13 (May 1986): 230–2. There is evidence to suggest that differences in generational experiences are particularly significant in rural America. On the basis of interviews with older residents of Union County, Illinois, anthropologist Jane Adams has found strikingly different outlooks on life between those who came of age during World War I, when the farm economy was booming, and those who came of age during the depression and World War II. Jane Adams, *The Transformation of Rural Life: Southern Illinois, 1890–1990* (Chapel Hill: University of North Carolina Press, 1994), p. 163.

18. Katherine Newman has found a generational divide between parents who came of age during the prosperous postwar period and children of the "baby boom," who now confront the economic uncertainty of the contemporary period. Postwar parents are hard-pressed to explain their children's lack of financial stability and material restraint, and privately worry that they were "spoiled" by the affluence of their youth. Unable to

achieve the same standard of living that their parents were able to provide, boomers feel deprived of the rewards that education and hard work were supposed to ensure. Katherine S. Newman, *Declining Fortunes: The Withering of the American Dream* (New York: Basic Books, 1993).

19. Jerome M. Stam, S. R. Koenig, S. E. Bentley, and H. F. Gale Jr., *Farm Financial Stress, Farm Exits, and Public Sector Assistance to the Farm Sector in the 1980s* (USDA: Economic Research Service, Agricultural Economic Report, no. 645, April 1991).

20. Linda M. Lobao and Paul Lasley, "Farm Restructuring and Crisis in the Heartland: An Introduction," in *Beyond the Amber Waves of Grain,* Lasley et al., pp. 1–27.

21. Patrick H. Mooney and Theo J. Majka, *Farmers' and Farm Workers' Movements: Social Protest in American Agriculture* (New York: Twayne, 1995), p. 106.

22. On the ideological continuity between historically distinct rural protest movements, see Catherine McNicol Stock, *Rural Radicals: Righteous Rage in the American Grain* (Ithaca, N.Y.: Cornell University Press, 1996); and Mary Summers, "The 1980s: What Still Stands? Farmers Movements Building the American State," in *The Countryside in the Age of the Modern State: Essays on the Political History of Modern America,* eds. Catherine McNicol Stock and Robert Johnston (Ithaca, N.Y.: Cornell University Press, forthcoming).

23. See, for example, Joshua Hammer, "A Double Slaying in Rural Minnesota Spotlights the Distress of America's Debt-Ridden Farmers," *People Magazine* (31 October 1983); Andrew H. Malcolm, "A Murder Testifies to the Death of Family Farms," *New York Times* (29 April 1984); Kevin Klose, "Killings, Suicide Echo Farm Crisis; Many Are in Financial Plight Blamed for Iowan's Rampage," *Washington Post* (11 December 1985); Andrew H. Malcolm, "Deaths in the Iowa Prairie: Four New Victims of Economy," *New York Times* (11 December 1985). Several journalists were inspired to write full-length books on these events: Bruce Brown, *Lone Tree: A True Story of Murder in America's Heartland* (New York: Crown Publishers, 1989); Andrew H. Malcolm, *Final Harvest: An American Tragedy* (New York: Times Books, 1986); James Corcoran, *Bitter Harvest: Gordon Kahl and the Posse Comitatus: Murder in the Heartland* (New York: Penguin, 1990); Osha Gray Davidson, *Broken Heartland: The Rise of America's Rural Ghetto*, exp. ed. (Iowa City: University of Iowa Press, 1996). These accounts explicitly assume a link between the farm crisis, political extremism, and a distressed farmer's murderous rampage. For a notable exception, see Joseph Amato's *When Father and Son Conspire: A Minnesota Farm Murder* (Ames: Iowa State University Press, 1988).

24. The classic discussion is Richard Hofstadler, *Age of Reform: From*

Bryan to FDR (New York: Vintage, 1955). See also, Richard Hofstadler, *The Paranoid Style in American Politics* (Chicago: University of Chicago Press, 1979 [1965]); Daniel Pipes, *Conspiracy: How the Paranoid Style Flourishes and Where it Comes From* (New York: Free Press, 1997); and Gordon Wood, "Conspiracy and the Paranoid Style," *William and Mary Quarterly* 3 (July 1982): 401–41.

25. Populist movements in the United States, particularly in the prairie states at the close of the nineteenth century, have inspired a rich and diverse body of historical analysis. Lawrence Goodwyn, *Populist Moment: A Short History of the Agrarian Revolt in America* (New York: Oxford University Press, 1978); Michael Kazin, *The Populist Persuasion: An American History* (New York: Basic Books, 1995); Robert McMath, *American Populism: A Social History* (New York: Hill and Wang, 1993); Jeff Ostler, *Prairie Populism: The Fate of Agrarian Radicalism in Kansas, Nebraska, and Iowa, 1880–1892* (Lawrence: University Press of Kansas, 1993); and C. Vann Woodward, *Tom Watson: Agrarian Rebel* (New York: Macmillan, 1938).

26. On the social history of farming in the United States, see Adams, *The Transformation of Rural Life*; John Mack Faragher, *Sugar Creek: Life on the Illinois Prairie* (New Haven: Yale University Press, 1986); Sally McMurry, *Transforming Rural Life: Dairying Families and Agricultural Change, 1820–1885* (Baltimore: Johns Hopkins University Press, 1995); and Mary Neth, *Preserving the Family Farm: Women, Community, and the Foundations of Agribusiness in the Midwest, 1900–1940* (Baltimore: Johns Hopkins University Press, 1995).

27. "Cannibalization" usually refers to the process by which larger, innovative farms are able to force smaller farms out of business, primarily because of the economic advantage that technological advancement is thought to confer. Willard W. Cochrane, *The Development of American Agriculture: A Historical Analysis* (Minneapolis: University of Minnesota Press, 1979), pp. 389–90. As Peggy F. Barlett has convincingly demonstrated, however, the farm crisis was not, in itself, a feeding frenzy, nor did it immediately result in one. In Dodge County, Georgia, where her research was conducted, there is little evidence that large farms acquired land through the failure of smaller farms between 1977 and 1986. Indeed, contrary to expectation, the farmers most likely to fail were the large-scale innovators, not the medium-sized farmers who exercise a "cautious" management style. Thus, Barlett concludes that there is "no support for the expectation of widespread cannibalism of medium-sized farms by large farms." *American Dreams, Rural Realities: Family Farms in Crisis* (Chapel Hill: University of North Carolina Press, 1993), pp. 217–18. While Barlett's point is well taken, the fact of economic competition between farmers remains—and the trend toward larger and fewer farms continues unabated.

28. James C. Scott, *Domination and the Arts of Resistance: Hidden Transcripts* (New Haven: Yale University Press, 1990).

29. Ibid., p. 20.

30. Marshall Sahlins has used this concept to argue that different cultures have distinctive modes of historical production—or as he puts it, "different cultures, different historicities"—a position with which I am in basic agreement. My point is that the culture of rural America is not fundamentally different from that of American society itself. Marshall Sahlins, *Islands of History* (Chicago: University of Chicago Press, 1985).

31. This self-positioning is particularly evident in the middle American belief that public services disproportionately favor the undertaxed rich and benefit the freeloading poor. Herbert Gans, *Middle American Individualism: The Future of Liberal Democracy* (New York: Free Press, 1988). For an illuminating look at how the feeling of being "squeezed" from above and below results in a sense of political and economic dispossession, see Jonathan Reider's *Canarsie: The Jews and Italians of Brooklyn against Liberalism* (Cambridge: Harvard University Press, 1985).

Chapter Two

1. Osha Gray Davidson offers a journalist's account of this period in *Broken Heartland: The Rise of America's Rural Ghetto*, exp. ed. (Iowa City: University of Iowa Press, 1996). For another journalist's personal story, see Howard Hohn, *The Last Farmer: An American Memoir* (New York: Harper and Row, 1988).

2. Anthropologist Peggy F. Barlett documents instances of excessive "generosity" on the part of the Farmers Home Administration between 1977 and 1982 in her study of rural Dodge County, Georgia. *American Dreams, Rural Realities: Family Farms in Crisis* (Chapel Hill: University of North Carolina Press, 1993), pp. 182–83. On the "economic logic" of farm expansion during this period, see William P. Browne, Jerry R. Skees, Louis E. Swanson, Paul B. Thompson, and Laurian J. Unnevehr, *Sacred Cows and Hot Potatoes: Agrarian Myths in Agricultural Policy* (Boulder, Colo.: Westview Press, 1992); Nora L. Brooks, Thomas A. Stucker, and Jennifer A. Bailey, "Income and Well-being of Farmers and the Farm Financial Crisis," *Rural Sociology* 51, no. 4 (1986): 391–405; and Neil E. Harl, *The Farm Debt Crisis of the 1980s* (Ames: Iowa State University, 1990).

3. Agriculture's first golden age occurred between 1900 and 1920. During this period, the value of the average farm more than tripled and real farm incomes (gross income adjusted for inflation) increased by 40 percent. See David B. Danbom, *Born in the Country: A History of Rural America* (Baltimore: Johns Hopkins University Press, 1995).

4. Kenneth L. Peoples, David Freshwater, Gregory D. Hanson, Paul T. Prentice, and Eric P. Thor, *Anatomy of an American Agricultural Credit Crisis: Farm Debt in the 1980s* (Lanham, M.D.: Rowman & Littlefield, 1992), p. 21.

5. Darwinian thinking becomes popular when it provide a cultural rationale for economic trends that demand "flexible" adaptations on the part of business firms and individual employees. In this rhetoric, for example, deindustrialization signals the evolutionary end of smokestack manufacturing and blue-collar labor. Kathryn Marie Dudley, *The End of the Line: Lost Jobs, New Lives in Postindustrial America* (Chicago: University of Chicago Press, 1994). See also Emily Martin, *Flexible Bodies: Tracking Immunity in American Culture from the Days of Polio to the Age of AIDS* (Boston: Beacon Press, 1994).

6. The University of Minnesota Extension Service has provided educational outreach to rural Minnesota since 1909. As part of the land grant system, the Minnesota Extension Service is integral to the network of research and educational agencies that formulates agricultural policy in the United States. Willard W. Cochrane, *The Development of American Agriculture: A Historical Analysis* (Minneapolis: University of Minnesota Press, 1979); Donald Paarlberg, *Farm and Food Policy* (Lincoln: University of Nebraska Press, 1980).

7. The Minnesota Farmer-Lender Mediation Act was passed in March 1986 as article 1 of Minnesota's 1986 Omnibus Farm Bill. The law requires creditors to mediate delinquent farm debts through the Farm Credit Mediation Program, administered by the University of Minnesota Extension Service. Mandatory farm credit mediation is intended to facilitate negotiation between borrowers and creditors by introducing a neutral party into the discussions. Mediators are community volunteers who complete Extension training in conflict resolution skills, farm mediation process, and farm finance issues. Kathy Mangum and Joyce Walker, *Mandatory Farm Credit Mediation: What It Is and How It Works* (Minnesota Extension Service, University of Minnesota, AD-FO-3008-S, rev. ed. 1993).

8. Farmers were not the only ones rushing out on a speculative limb. For a look at financial excess elsewhere in America at that time, see James R. Bath, *The Great Savings and Loan Debacle* (Washington, D.C.: American Enterprise Institute, 1991); Michael Lewis, *Liar's Poker: Rising through the Wreckage on Wall Street* (New York: Norton, 1989); Martin E. Lowy, *High Rollers: Inside the Savings and Loan Debacle* (New York: Praeger Group, 1992); and Mary Zey, *Banking on Fraud: Drexel, Junk Bonds, and Buyouts* (New York: Aldine deGruyter, 1993).

9. Peoples et al., *Anatomy of an American Agricultural Credit Crisis*, pp. 32–33.

10. Among analysts of American trade policy, there appears to be some confusion about the economic consequences of Carter's grain embargo. Compare, for example, Anthony Campagna's discussion of the embargo from the perspective of the Reagan presidency versus that of the Carter administration. In *The Economy in the Reagan Years: The Economic Consequences of the Reagan Administration* (Westport, Conn.: Greenwood Press, 1994), Campagna writes: "Agricultural export markets [that farmers had been promised] did not develop, and, in fact, the US lost markets in the USSR food embargo fiasco" (p. 177). A year later, in *Economic Policy in the Carter Administration* (Westport, Conn.: Greenwood Press, 1995), he writes: "In 1980, agricultural exports increased by 18 percent, despite the embargo placed on the USSR by the administration following its invasion of Afghanistan. The loss of exports to the Soviet Union was made up by sales to Asia, Africa, and Latin America, and to Western Europe, which had been experiencing poor harvests" (p. 123).

11. Herve Varenne, *Americans Together: Structured Diversity in a Midwestern Town* (New York: Teachers College Press, 1977), pp. 92–95.

12. Jerome M. Stam, S. R. Koenig, S. E. Bentley, and H. F. Gale Jr., *Farm Financial Stress, Farm Exits, and Public Sector Assistance to the Farm Sector in the 1980s* (USDA: Economic Research Service, Agricultural Economic Report, no. 645, April 1991), p. 39.

13. Exploring these other worlds is what I take to be the central aim of anthropology, which is, as Clifford Geertz has put it, "the enlargement of the universe of human discourse." *The Interpretation of Cultures* (New York: Basic Books, 1973), p. 14. Such a project also involves recognizing that the individualistic model of the person in economics and in the social sciences is itself a cultural construction. As Mary Douglas and Steven Ney cogently argue, anthropology's multicultural perspective demands that we call this concept of personhood into question. *Missing Persons: A Critique of Personhood in the Social Sciences* (Berkeley: University of California Press, 1998). My critique of public discourse in American society owes much to the theoretical framework developed by Mary Douglas, especially, *Risk and Blame: Essays in Cultural Theory* (London: Routledge, 1992); and the substantive focus on ritual and experience pioneered by Victor Turner, particularly in *From Ritual to Theatre* (New York: Performing Arts Journal Press, 1982); and *The Anthropology of Experience*, eds. Victor W. Turner and Edward M. Bruner (Urbana: University of Illinois Press, 1986).

Chapter Three

1. James C. Scott, *The Arts of Domination and Resistance: Hidden Transcripts* (New Haven: Yale University Press, 1990).

2. Life insurance companies held 25.2 percent of national farm real es-

tate loans in 1957. Their market share declined to 11.4 percent in 1984, largely due to competition from the Federal Land Banks. David Lins, "Life Insurance Company Lending to Agriculture," *Agricultural Finance Review* 41 (July 1981): 41–49; and Patrick Mooney, *My Own Boss?: Class, Rationality, and the Family Farm* (Boulder, Colo.: Westview Press, 1988), pp. 133–34.

3. Kenneth L. Peoples, David Freshwater, Gregory D. Hanson, Paul T. Prentice, and Eric P. Thor, *Anatomy of an American Agricultural Credit Crisis: Farm Debt in the 1980s* (Lanham, M.D.: Rowman & Littlefield, 1992), fig. 1.3, p. 13.

4. The depression-era programs were called the Resettlement Administration and the Farm Security Administration, respectively. Farmers Home was commonly known as FHA until 1974, when the USDA adopted the official abbreviation of FmHA to distinguish Farmers Home from other federal agencies with the same initials. Farmers, however, continue to use the acronym FHA when speaking of the agency. U.S. Department of Agriculture, *A Brief History of Farmers Home Administration: Serving Rural America for 50 Years* (USDA: Farmers Home Administration, February 1986). In December 1994, USDA restructuring gave the agency a new name—Rural Economic and Community Services—and FmHA's farm loan portfolios were transferred to what is now called the Farm Service Agency.

5. Since the major portion of my fieldwork was completed in 1994, before the USDA restructuring and official name change, I exercise the ethnographer's prerogative and continue to refer to Farmers Home in the present tense.

6. Star Prairie was not the only county hit by the blizzard of 1940. For an arresting account of this statewide disaster based on oral history interviews, see William H. Hull, *All Hell Broke Loose: Experiences of Young People during the Armistice Day 1940 Blizzard* (Edina, Minn.: Stanton Publication Service, 1996 [1985]).

7. Bourdieu argues that there are different types of capital—different ways, that is, of appropriating social energy or accumulated labor. In addition to the material or "economic" form of capital—which can be immediately and directly exchanged for money—Bourdieu identifies two immaterial forms of capital: "cultural capital" and "social capital." Cultural capital is the knowledges or competencies that members of a society acquire through education, training, and experience. Social capital is the network of interpersonal connections that individuals acquire through the exchange of goods and services. Bourdieu's key insight is that these different types of capital are interchangeable—that is, with more or less difficulty, one type can be converted into another. Pierre Bourdieu, "The Forms of Capital," in *Handbook of Theory and Research in the Sociology*

of Education, ed. John G. Richardson (New York: Greenwood, 1986), pp. 241–58. For a discussion of the convertibility of different forms of capital, see Craig Calhoun, "Habitus, Field, and Capital: The Question of Historical Specificity," in *Bourdieu: Critical Perspectives,* eds. Craig Calhoun, Edward LiPuma, and Moishe Postone (Chicago: University of Chicago Press, 1993).

8. Pierre Bourdieu has given the term "habitus" to cultural capital that assumes an "embodied" form. *Outline of a Theory of Practice,* trans. R. Nice (Cambridge: Cambridge University Press, 1977 [1972]), pp. 78–87.

9. Catherine McNicol Stock, *Main Street in Crisis: The Great Depression and the Old Middle Class on the Northern Plains* (Chapel Hill: University of North Carolina Press, 1992), pp. 87, 99. See also Jean-Christophe Agnew, "A Touch of Class," *Democracy* 3 (spring 1983): 59–72; and Warren Susman, *Culture as History: The Transformation of American Society in the Twentieth Century* (New York: Pantheon Books, 1984). As Jane Adams has observed, rural areas underwent a dramatic transformation in the years between the beginning of the New Deal and the end of World War II. New Deal progressives, like the postwar economic planners who succeeded them, believed that the only way to make farm incomes rise was to reduce the total number of farmers. This could be accomplished, they agreed, by encouraging technological innovation and the out-migration of "inefficient" farmers. Jane Adams, *The Transformation of Rural Life: Southern Illinois, 1890–1990* (Chapel Hill: University of North Carolina Press, 1994), pp. 160–61, 183.

10. The new character of FmHA supervisors appears to parallel the change in management that accompanies restructuring in other industries. Katherine Newman found a similar shift in workplace culture when the Singer factory in Elizabeth, New Jersey, began recruiting executives from unrelated businesses to replace the older "sewing machine men" who traditionally supervised workers on the shop floor. Katherine S. Newman, "Turning Your Back on Tradition: Symbolic Analysis and Moral Critique in a Plant Shutdown," *Urban Anthropology* 14, nos. 1–3 (1985): 109–150.

11. Emanuel Melichar, "A Financial Perspective on Agriculture," *Federal Reserve Bulletin* 70 (January 1984): 1–13.

12. Neil E. Harl, *The Farm Debt Crisis of the 1980s* (Ames: Iowa State University Press, 1990).

13. The classic study is E. P. Thompson's "The Moral Economy of the English Crowd in the Eighteenth Century," *Past & Present* 50 (February 1971): 76–136. James C. Scott demonstrates the power of this approach in *The Moral Economy of the Peasant: Rebellion and Subsistence in Southeast Asia* (New Haven: Yale University Press, 1976).

Chapter Four

1. Middle-class wage earners are generally reluctant to mix money and family. Even when families fall upon hard times, considerable ambiguity surrounds the norms of obligation and exchange, such that even well-intended assistance can be taken as an affront and the direst of straits can be dismissed as the private problem of the relatives in need. Katherine S. Newman, *Falling from Grace: Downward Mobility in the Age of Affluence* (Berkeley: University of California Press, 1999 [1988]).

2. The viewing of old Westerns is a major motif in Jeanne Jordan and Steve Ascher's documentary *Troublesome Creek: A Midwestern*. The film chronicles the response of Jeanne's parents, Russ and Mary Jane Jordan, to the forced sale of the family farm in 1990. The significance of the Western and the moral struggle it encodes were evidently lost on New York critic Walter Goodman, who wrote of the film: "It's sad that Russ has to sell off his cows and Mary Jane has to sell off her beloved Ethan Allen furniture, but such is the price of a failed speculation, and the couple winds up in fair comfort. The producers' insistence on the elder Jordans watching Hollywood westerns on television ('Here are my parents, Russel and Mary Jane Jordan, doing what they like to do best: watching a good western') is a documentarian's conceit that quickly becomes annoying." Walter Goodman, "For a Farm, No Happy Ending," *New York Times* (14 April 1997), sec. C, p. 16.

3. Kenneth L. Peoples, David Freshwater, Gregory D. Hanson, Paul T. Prentice, and Eric P. Thor, *Anatomy of an American Agricultural Credit Crisis: Farm Debt in the 1980s* (Lanham, Md.: Rowman & Littlefield, 1992), pp. 38–41.

4. Michael Herzfeld, *The Social Production of Indifference: Exploring the Symbolic Roots of Western Bureaucracy* (Chicago: University of Chicago Press, 1992), p. 80.

5. Philip Van Hoff's foreclosure also took place before the Homesite Protection Act was passed as part of Minnesota's 1986 Omnibus Farm Bill. This law would have allowed him to repurchase his house and homesite after foreclosure.

6. For an excellent analysis of the factors that affect farm succession between father and son, and the inheritance of the patrimony by siblings, see Sonya Salamon, *Prairie Patrimony: Family, Farming, and Community in the Midwest* (Chapel Hill: University of North Carolina Press, 1992).

7. I am drawing on Erving Goffman's classic definition of a "primal scene" in sociology. As Goffman writes: "Stigma involves not so much a set of concrete individuals who can be separated into two piles, the stigmatized and the normal, as a pervasive two-role process in which every individual participates in both roles, at least in some connections and in

some phases of life. The normal and the stigmatized are not persons but rather perspectives." *Stigma: Notes on the Management of a Spoiled Identity* (New York: Simon and Schuster, 1963), pp. 137–38.

Chapter Five

1. The lawsuit was brought against the secretary of agriculture—Richard Lyng, and later, John Block—by Dwight Coleman and eight other North Dakota farmers. The lawsuit challenged the agency's foreclosure and liquidation procedures, charging (1) that they did not permit loan deferrals in cases where farmers were in default due to forces beyond their control, and (2) that they violated farmers' right to due process by cutting off their income when a loan was called due. In this so-called "starve out" tactic, the agency refused to release a farmer's production income—which came in checks made out to FmHA—claiming this money as payment against delinquent loan accounts. In essence, as farm activists observed at the time, the lawsuit accused the federal government of abandoning its mission of "fostering and encouraging the family farm system." "North Dakota Farmer's Suit Widened to Include 44 States," *New York Times* (29 October 1983), sec. 1, p. 6.

2. Federal Judge Bruce Van Sickle of Bismarck, North Dakota, ruled in favor of the plaintiffs in November 1983. "Around the Nation, US Court Restricts Farm Foreclosures," *New York Times* (15 November 1983), sec. A, p. 16. In 1984 Van Sickle entered an injunction in the case that stopped FmHA from liquidating or foreclosing on borrowers until the secretary of agriculture implemented a program allowing farmers to apply for deferrals of principal and interest payments when they lost income because of unforeseen circumstances. "New Farm Foreclosure Rules," *New York Times* (1 January 1984), sec. A, p. 14. In November 1985, the Farmers Home Administration published regulations that established a loan deferral program, including procedures for notifying farmers of foreclosures. Farmers in the Coleman case complained that the notification procedures didn't give borrowers enough information about their situation and didn't offer an adequate opportunity to apply for agency programs that might prevent foreclosure. Van Sickle again ordered the FmHA to halt foreclosures and liquidations until it revised its notification methods. In 1986 the Farmers Legal Action Group (FLAG) again filed a plea for federal relief, arguing that farmers were being liquidated because FmHA had disregarded the judge's orders. "Farmers Act to Block New US Laws on Loans," *New York Times* (24 January 1986), sec. A, p. 12. In May 1987 Van Sickle ordered that all adverse actions, including foreclosures, again be stayed. He ruled that FmHA notices of intent to take adverse action were unconstitutional. The action immediately affected 78,000 farmers allowing them to use $714 million in money they had earned through farm

production through its release to them for living and operating expenses. "Farmers Win on Foreclosures," *New York Times* (4 June 1987), sec. A, p. 25. Nonetheless, FmHA was again charged with circumventing the law by accelerating loans and refusing to release living and operating expenses. In March 1988 Van Sickle ordered that the FmHA give written notice and allow for an appeals hearing when deciding not to release a borrower's production income for payment of necessary living and operating expenses.

3. Kenneth L. Peoples, David Freshwater, Gregory D. Hanson, Paul T. Prentice, and Eric T. Thor, *Anatomy of an American Agricultural Credit Crisis: Farm Debt in the 1980s* (Lanham, M.D.: Rowman & Littlefield, 1992); Jerome M. Stam, Steven R. Koenig, Susan E. Bently, H. Frederick Gale Jr., *Farm Financial Stress, Farm Exits, and Public Sector Assistance to the Farm Sector in the 1980s* (USDA: Economic Resource Service Agricultural Economic Report, no. 645, April 1991).

4. Reverend David Ostendorf was the executive director of the Iowa-based PrairieFire at its founding in 1985. He recalls of this period: "I will never forget going into the office on the second day of January in 1985. I got in early, 7:00 to 7:30 that morning, and our phones were ringing incessantly at that hour of the morning. To this day, I still look back and say, 'What was it on that day that caused the logjam to break?' I guess it was a lot of farmers after the holidays, they knew they had notes due and so on. But that was the day that things really began to come together. From that day forward through the spring, it was an absolutely endless scene of protests, of meetings, of rallies, all of the eruption occurred. Media coverage helped us build the movement too. We stopped sales, and that gave people a sense that we could do some things. But every week almost, there was another protest of some kind or other. It was happening all over the region" (Interview with the author, Chicago, Ill., 17 November 1995).

5. Based on a statewide survey of Iowa farmers, Gordon Bultena, Paul Lasley, and Jack Geller concluded that the farmers at greatest risk of failure were younger, better educated, large-scale operators. "The Farm Crisis: Patterns and Impacts of Financial Distress Among Iowa Farm Families," *Rural Sociology* 51, no. 4 (1986): 436–48. Economist Neil Harl offers a comprehensive overview of the risk factors involved in *The Farm Debt Crisis of the 1980s* (Ames: Iowa State University Press, 1990); as do sociologists Steve H. Murdock, Don E. Albrecht, Rita R. Hamm, F. Larry Leistritz, and Arlen G. Leholm, "The Farm Crisis in the Great Plains: Implications for Theory and Policy Development," *Rural Sociology* 51, no. 4 (1986): 406–435; and Steve H. Murdock and F. Larry Leistritz, eds., *The Farm Financial Crisis: Socioeconomic Dimensions and Implications for Producers and Rural Areas* (Boulder, Colo.: Westview, 1988).

6. Alfred Reginald Radcliffe-Brown, "On Joking Relationships,"

Africa 13, no. 3 (1940): 195–210; and *Structure and Function in Primitive Society* (Glencoe, Ill.: Free Press, 1959), pp. 90–115. Radcliffe-Brown argues that ritualized joking behavior works to resolve, or smooth over, structural tensions that arise in social relationships, such as those created by marriage in a kin-based society. In a matrilineal kinship system, for example, the authority that men have over their sister's sons is necessarily limited, since these young men will eventually marry into a different matrilineage. In an important contribution to this general theory, Mary Douglas points out that jokes may also challenge and disrupt the social structure by giving voice to its inconsistencies. "The Social Control of Cognitions: Some Factors in Joke Perception," *Man* 3 (1968): 361–76.

7. Linda M. Lobao, "Organizational, Community, and Political Involvement as Responses to Rural Restructuring," in *Beyond the Amber Waves of Grain: An Examination of Social and Economic Restructuring in the Heartland,* eds. Paul Lasley, F. Larry Leistritz, Linda M. Lobao, Katherine Meyer (Boulder, Colo.: Westview Press, 1995), pp. 183–205.

8. Federal commodity programs aim to reduce surpluses—and thereby stabilize prices—by paying farmers not to produce specific commodities. Payments are linked to a farmer's level of production. In 1988 farms with at least $250,000 in sales constituted about 12 percent of all participating farms but received 34 percent of total payments. G. Whittaker, "Payments Go to Largest Farms," *Agricultural Outlook* (U.S. Department of Agriculture, Economic Resource Service Agricultural Economic Report, April 1990), pp. 25–27; and Stam et al., *Farm Financial Stress, Farm Exits, and Public Sector Assistance,* p. 37.

9. A boundary crisis does not necessarily mean that a structural change has occurred within the community itself, but that a preexisting source of social division has been contested and needs to be more precisely defined and actively defended. Erikson writes: "Single encounters between the deviant and his community are only fragments of an ongoing social process. Like an article of common law, boundaries remain a meaningful point of reference only so long as they are repeatedly tested by persons on the fringes of the group and repeatedly defended by persons chosen to represent the group's inner morality. . . . Deviant forms of behavior, by marking the outer edges of group life, give the inner structure its special character and thus supply the framework within which the people of the group develop an orderly sense of their own cultural identity." Kai T. Erikson, *Wayward Puritans: A Study in the Sociology of Deviance* (New York: Wiley, 1966), pp. 12–13.

10. Thanks in large part to the efforts of country-western singer Willie Nelson, a September 1985 Farm Aid concert raised over $9 million to aid struggling American farmers. Although the proceeds fell far short of the

$50 million organizers expected, the event succeeded in drawing national attention and sympathy to the economic plight of American farmers. "Nelson Lobbies for Farm Aid," *Los Angeles Times* (21 September 1985), sec. 5, p. 1; "Concert Money to Aid Farmers in 17 States," *New York Times* (29 September 1985), sec. 1, p. 6; "Farm Couple Declines Outpouring of Aid," *Washington Post* (29 September 1985), A18.

Chapter Six

1. The distinction between "front stage" and "backstage" impression management is Erving Goffman's, in *The Presentation of Self in Everyday Life* (Garden City, N.Y.: Doubleday, 1959).

2. The Thompson brothers are perfect examples of the kind of farmer who was most likely to survive the farm crisis nationwide. As anthropologists have discovered, there is a recognizable difference between cautious and ambitious farm management styles—and a strong tendency for more ambitious managers to be at greater risk of failure. In general, this difference is thought to emanate from an underlying cultural difference in the way farmers define success and, hence, from the business strategies they employ to achieve it. In the typology proposed by Sonya Salamon, for example, "cautious" managers are those who define success in terms of family continuity. They value keeping the farm in the family and therefore attempt to minimize financial risk by avoiding debt, even if this means settling for a modest standard of living. "Ambitious" managers, in contrast, are those who define success as maximizing financial profit. They value personal achievement for themselves and upward mobility for their children, and therefore do not hesitate to expand their operation through debt financing, even if this means putting the farm itself at risk. Salamon links these divergent management styles to ethnic identity, suggesting that certain ethnic groups—in this case, German "yeomen" farmers—inculcate family values that allow them to resist the "entrepreneurial" ethos of Yankee families or mixed-ethnic groups. Sonya Salamon and Karen-Davis Brown, "Middle-Range Farmers Persisting through the Agricultural Crisis," *Rural Sociology* 51, no. 4 (1986): 503–12; Salamon, "Ethnic Determinants of Farm Community Character," in *Farm Work and Fieldwork: American Agriculture in Anthropological Perspective*, ed. Michael Chibnick (Ithaca: Cornell University Press, 1987), pp. 167–88; and *Prairie Patrimony: Family, Farming, and Community in the Midwest* (Chapel Hill: University of North Carolina Press, 1992). Along similar lines, Peggy F. Barlett argues that rural communities are increasingly unable to resist the industrial values of the wider society. She attributes this "erosion of agrarian values" to persistent rural poverty, migration to urban areas, and the mixture of ethnic groups. Barlett, "The Crisis in Family

Farming: Who Will Survive?" in *Farm Work and Fieldwork*, ed. Chibnick, pp. 29–57; and *American Dreams, Rural Realities: Family Farms in Crisis* (Chapel Hill: University of North Carolina Press, 1993). Regardless of how the distinction between cautious and ambitious cultural orientations is figured—as yeoman versus entrepreneur or agrarian versus industrial—the ethnographic research is unequivocal on one point: whereas the entrepreneurial style was strongly encouraged during the 1970s, in retrospect it is the cautious farmer who has endured. While not inconsistent with this research, my interviews suggest a different interpretation. As Virgil Thompson and other Star Prairie farmers attest, an entrepreneurial ethos can coexist with—and reinforce—what I am calling the community's "code of frugality."

3. For the classic analysis of agriculture's "technology treadmill," see Willard W. Cochrane, *The Development of American Agriculture: A Historical Analysis* (Minneapolis: University of Minnesota Press, 1979). Yet we would do well to remember, as Jane Adams points out, that the apparent "inevitability" of technological progress and the particular course it takes are as much a product of economic and governmental policies as they are of specific forms of "resistance to modernity." *The Transformation of Rural Life: Southern Illinois, 1890–1990* (Chapel Hill: University of North Carolina Press, 1994), p. 182; and "Resistance to 'Modernity': Southern Illinois Farm Women and the Cult of Domesticity," *American Ethnologist* 20, no. 1 (1993): 89–113.

4. Katherine S. Newman, *Falling from Grace: Downward Mobility in the Age of Affluence* (Berkeley: University of California Press, 1999 [1988]).

5. Clearly, Bea Hagendorf's social activities support the family's farm in ways that are difficult to measure economically. Deborah Fink has illuminated the social history of women's contributions to agriculture in *Open Country, Iowa: Rural Women, Tradition and Change* (Albany: State University of New York Press, 1986); and *Agrarian Women: Wives and Mothers in Rural Nebraska, 1880–1940* (Chapel Hill: University of North Carolina Press, 1992); as has Katherine Jellison in *Entitled to Power: Farm Women and Technology, 1913–1963* (Chapel Hill: University of North Carolina Press, 1993).

6. I am indebted to Jean-Christophe Agnew for the insight that farmers' consumer culture is as much about "conspicuous production" as it is about conspicuous consumption. A new pickup truck raises questions not only about "showing off," but also about the claims to social advancement associated with entrepreneurship. See John Brooks, *Showing Off in America: From Conspicuous Consumption to Parody Display* (Boston: Little, Brown and Company, 1981 [1979]); and Donald Finlay Davis,

Conspicuous Production: Automobiles and Elites in Detroit, 1899–1933 (Philadelphia: Temple University Press, 1988).

7. Erving Goffman, *Stigma: Notes on the Management of Spoiled Identity* (New York: Simon and Schuster, 1963).

8. Eve Kosofsky Sedgwick, *The Epistemology of the Closet* (Berkeley: University of California Press, 1990), p. 67; and D. A. Miller, *The Novel and the Police* (Berkeley: University of California Press, 1988).

9. Michael J. Belyea and Linda M. Lobao, "Psycho-Social Consequences of Agricultural Transformation: The Farm Crisis and Depression," *Rural Sociology* 55, no. 1 (spring 1990); Paul Rosenblatt, *Farming Is in Our Blood: Farm Families in Economic Crisis* (Ames: Iowa State University Press, 1990); and Mary Van Hook, "Family Response to the Farm Crisis: A Study in Coping," *Social Work* 35, no. 5 (September 1990).

10. David Celani, *The Illusion of Love: Why the Battered Woman Returns to Her Abuser* (New York: Columbia University Press, 1994); Donald Dutton, *The Domestic Assault of Women: Psychological and Criminal Justice Perspectives* (Vancouver: University of British Columbia, 1995); and Judith Herman, *Trauma and Recovery* (New York: Basic Books, 1992).

Chapter Seven

1. The auction is a ritualized way of stripping farmers of their occupational and class status, much as a plant closing in a deindustrializing economy serves to expel factory workers from the middle class. I introduce this line of analysis in *The End of the Line: Lost Jobs, New Lives in Postindustrial America* (Chicago: University of Chicago Press, 1994).

2. Victor Turner, *The Ritual Process: Structure and Anti-structure* (Ithaca: Cornell University Press, 1969), p. 95.

3. In a medical treatment manual for general practitioners, Kenneth Hambly and Alice Jane Muir offer a widely accepted definition of stress: "Stress is a condition in which there is a maladaptively high level of adrenergic arousal, which may be acute and/or chronic, resulting in a range of unpleasant physical, psychological and behavioral problems." *Stress Management in Primary Care* (Oxford: Butterworth-Henemann, 1997), p. 6. Stress can only be reduced or managed, not "cured," since virtually every aspect of modern life is thought to be potentially stressful. Stress can be diagnosed in the family, personal relationships, social conditions, and the workplace. In such a model, there is no meaningful difference between the unpleasant effects of raising a toddler, combating racism, living in poverty, or putting in long hours on the job. See also Cary L. Cooper, *The Stress Check: Coping with the Stresses of Life and Work* (Englewood Cliffs, N.J.: Prentice Hall, 1981).

4. To one degree or another, it should be noted, all social interactions partake of this "as if" quality. The identities we have in public largely depend on the social category and attributes that others impute to people with our characteristics. Erving Goffman has called this a "virtual" social identity—a characterization made "in effect"—to distinguish it from the "actual" social identity we have when our idiosyncratic attributes become known. *Stigma: Notes on the Management of Spoiled Identity* (New York: Simon and Schuster, 1963), pp. 2–3.

5. These accounts of economic loss bear a striking resemblance to passages in the Judeo-Christian scriptures where God is depicted as wounding the human body. As literary scholar Elaine Scarry observes, "scenes of wounding" serve to account for and demonstrate "the power and perfectibility of the divine and the imperfection and vulnerability of the human." The God who causes plagues, floods, fires, locusts, leprous sores, disease, hunger, and so forth is figured as the preeminent power, capable of affecting alterations in the human body, but is immune from such alterations Himself. Elaine Scarry, *The Body in Pain: The Making and Unmaking of the World* (New York: Oxford University Press, 1985), p. 183. On the loss of voice in suffering, see David B. Morris, "About Suffering: Voice, Genre, and Moral Community," in *Social Suffering*, eds. Arthur Kleinman, Veena Das, and Margaret Lock (Berkeley: University of California Press, 1997), pp. 25–45.

6. Judith Butler, *Excitable Speech: A Politics of the Performative* (New York: Routledge, 1997), pp. 36–37.

7. Of all the legislative reactions to the farm crisis, few have had as lasting an impact on the experience of farm loss as homestead protection acts. By giving families the option to buy their farm residence after foreclosure, states took a welcome stand against the forces draining rural America of productive young families. Not that this required great economic sacrifice, however. In an era of agricultural consolidation, few buyers of farmland have an interest in the homesites that dot the landscape. With tax laws that render these structures a liability, more easily razed than rented, those in a position to acquire their neighbor's land are generally relieved to have an unproductive farmhouse, often surrounded by other buildings and a windbreak, carved out of their purchase. The flip side of this bargain is the emotional toll it can take on the families who remain, always reminded of their loss by the sight of someone else farming their land. As sociologist Peter Marris has persuasively argued, the attachment we feel to our work and places of employment is powerfully rooted in the structure of our economic institutions as well as in our individual psychosocial development. *Loss and Change* (New York: Pantheon, 1974); and *The Politics of Uncertainty: Attachment in Private and Public Life* (London: Routledge, 1996).

Chapter Eight

1. In *America's Working Man* (Chicago: University of Chicago Press, 1984), David Halle makes the point that production workers can identify as "middle" class, even though they think of themselves as "working" men on the job.

2. Paul C. Rosenblatt also notes the prevalence of this expression among Minnesota farmers in *Farming Is in Our Blood: Farm Families in Economic Crisis* (Ames: Iowa State University Press, 1990).

3. For an overview of rural activism and farm movements, see Catherine McNicol Stock, *Rural Radicals: Righteous Rage in the American Grain* (Ithaca, N.Y.: Cornell University Press, 1996); and Patrick H. Mooney and Theo J. Majka, *Farmers' and Farm Workers' Movements: Social Protest in American Agriculture* (New York: Twayne Publishers, 1995).

4. David B. Danbom, *Born in the Country: A History of Rural America* (Baltimore: Johns Hopkins University Press, 1995), p. 190.

5. Ibid., p. 263. See also, Gregory D. Hanson, G. Hossein Parandvash, and James Ryan, *Loan Repayment Problems of Farmers in the Mid-1980s* (USDA: Economic Resource Service Agricultural Economic Report, no. 649, September 1991).

6. F. Larry Leistritz and Freddie L. Barnard, "Financial Characteristics of Farm Operators," in *Beyond the Amber Waves of Grain: An Examination of Social and Economic Restructuring in the Heartland*, eds. Paul Lasley, F. Larry Leistritz, Linda M. Lobao, and Katherine Meyer (Boulder, Colo.: Westview Press, 1995), p. 57. The researchers define net family income as "the return from farming to unpaid operator and family labor, management, and equity capital plus income from off-farm employment and other non-farm sources."

7. Data from the U.S. Department of Agriculture and U.S. Department of Commerce, Bureau of the Census, cited in *Anatomy of an American Agricultural Credit Crisis: Farm Debt in the 1980s,* eds. Kenneth L. Peoples, David Freshwater, Gregory D. Hanson, Paul T. Prentice, and Eric T. Thor (Lanham, M.D.: Rowman & Littlefield, 1992), p. 66.

8. Interview with the author, Chicago, Ill., 17 November 1995. David L. Ostendorf has published widely on issues related to the farm crisis, rural communities, and Christian theology. See, for instance, "Iowa's Rural Crisis of the 1980s: Of Devastation and Democracy," in *Family Reunions: Essays on Iowa*, ed. Thomas J. Morain (Ames: Iowa State University Press, 1995); with Dixon Terry, "Toward a Democratic Community of Communities: Creating a New Future with Agriculture and Rural America," in *Environmental Justice: Issues, Policies, and Solutions*, ed. Bunyan Bryant (Washington, D.C.: Island Press, 1994); and "A Protestant, Populist View," in *Is There a Moral Obligation to Save the Family Farm?* ed. Gary Comstock (Ames: Iowa State University Press, 1987).

9. Thomas A. Lyson argues that although opinion polls in 1986 showed support for federal aid to farmers, the lack of public outcry could be attributed to the facts that most Americans (1) have no direct contact with farming, (2) see no relationship between the cost of food and farm income or the number of farms, and (3) are inured to news reports that consistently portray farmers in economic crisis. "Who Cares about the Farmer? Apathy and the Current Farm Crisis," *Rural Sociology* 51, no. 4 (1986): 490–502.

10. As William P. Browne, Jerry R. Skees, Louis E. Swanson, Paul B. Thompson, and Laurian J. Unnevehr put it, "The operation of a competitive market, when farmers get what they want from lobbying, prevents higher prices from providing higher incomes in the long run." *Sacred Cows and Hot Potatoes: Agrarian Myths and Agricultural Policy* (Boulder, Colo.: Westview Press, 1992), p. 64.

11. This notion of community is consistent with Benedict Anderson's analysis of the nation as an "imagined community" in which the dead and the yet unborn are linked by narratives of a common destiny. *Imagined Communities: Reflections on the Origin and Spread of Nationalism,* 2nd ed. (London: Verso, 1992 [1983]).

12. There is a vast literature on individualism and community in America. Key works in the sociological tradition include Alexis de Tocqueville, *Democracy in America*, vols. 1 and 2 (New York: Vintage Books, 1945); David Riesman, *Lonely Crowd: A Study of the Changing American Character* (New Haven, Conn.: Yale University Press, 1961); Robert Bellah, Richard Madsen, William Sullivan, Ann Swidler, and Steve Tipton, *Habits of the Heart: Individualism and Commitment in American Life*, rev. ed. (Berkeley: University of California Press, 1996 [1985]); Philip Selznick, *The Moral Commonwealth: Social Theory and the Promise of Community* (Berkeley: University of California, 1992).

13. The impulse to treat value orientations as mutually exclusive positions or preferences oversimplifies what is, from another angle, meaningful cultural complexity. What appears as a tidy "opposition" to the moral philosopher is better treated, Kai Erikson has proposed, as an "axis of variation" that serves to organize cultural diversity. *Everything in Its Path: Destruction of Community in the Buffalo Creek Flood* (New York: Simon and Schuster), p. 82. For more on the so-called "communitarian critique of liberalism," see Alasdair MacIntyre, *After Virtue*, 2nd ed. (Notre Dame, Ind.: University of Notre Dame Press, 1984); Stephen Mulhall and Adam Swift, *Liberals and Communitarians* (Oxford: Blackwell, 1992); and Nancy L. Rosenblum, *Liberalism and the Moral Life* (Cambridge: Harvard University Press, 1989).

14. Even during the early decades of settlement, prairie communities were characterized by opposing impulses: the appeal of putting down

roots versus the lure of the open frontier. With local populations in constant flux and mercantile capitalism in progressive competition with neighborly barter systems, rural America during the pioneer period was also a world beset by contradictions. Yet as historian John Mack Faragher has observed, the development of a community culture was not antithetical to individualism or entrepreneurial pursuits: "Community did not 'break down' with the approach of the modern world; community, in fact, provided a means of making the transition to it." *Sugar Creek, Life on the Illinois Prairie* (New Haven: Yale University Press, 1986), p. 237.

15. On the transformation of the rural landscape, see Janet M. Fitchen, *Endangered Spaces, Enduring Places: Change, Identity, and Survival in Rural America* (Boulder, Colo.: Westview Press, 1991); and Caroline S. Tauxe, *Farms, Mines, and Main Streets: Uneven Development in a Dakota County* (Philadelphia: Temple University Press, 1993). On what it means to be "haunted" as a politically engaged scholar, see Avery Gorden, *Ghostly Matters: Haunting and the Sociological Imagination* (Minneapolis: University of Minnesota Press, 1997); and Jacques Derrida, *The Specters of Marx: The State of the Debt, the Work of Mourning, and the New International*, trans. Peggy Kamuf (New York: Routledge, 1994). Kathleen Stewart has written an innovative ethnography of a "ghostly landscape" in her study of the Appalachian poor who dwell amidst the ruins of West Virginia coal camps. *A Space on the Side of the Road: Cultural Poetics in an "Other" America* (Princeton: Princeton University Press, 1996).

16. Michael Adas, *Prophets of Rebellion: Millenarian Protest Movements against the European Colonial Order* (Cambridge: Cambridge University Press, 1979); James Mooney, *The Ghost-Dance Religion and the Sioux Outbreak of 1890* (Chicago: University of Chicago Press, 1965).

17. Kai Erikson, *Everything in Its Path,* pp. 153–54; and *A New Species of Trouble: Explorations in Disaster, Trauma, and Community* (New York: W. W. Norton, 1994).

INDEX